VEDIC CHRONOLOGY
AND
VEDANGA JYOTISHA

VEDIC CHRONOLOGY
AND
VEDANGA JYOTISHA

BY

LOKAMANYA

BAL GANGADHAR TILAK, B. A., LL. B.

GYAN

Published by

Gyan Publishing House
5, Ansari Road
Daryaganj, New Delhi-110002
Phone: 011-47034999, 9811692060
E-mail: books@gyanbooks.com

Distribution Network
gyanbooks.com
India, USA, Canada, UK, Australia, France

ISBN: 978-81-212-7963-5 (PB)
First Published, 1925

2nd Impression 2023

Printed at: Gyan Press, Delhi.

VEDIC CHRONOLOGY AND VEDANGA JYOTISHA
Author: LOKAMANYA BAL GANGADHAR TILAK

VEDIC CHRONOLOGY

AND

VEDANGA JYOTISHA.

[Containing also Chaldean and Indian Vedas and other miscellaneous essays.]

BY

LOKAMANYA

BAL GANGADHAR TILAK, B. A., LL. B.

Author of the "*Orion*" or "*Researches into the Antiquity of the Vedas*", "*The Arctic Home in the Vedas*", "*Gitâ-Rahasya*" &c. &c.

CONTENTS.

Scheme of Trans-literation.

अ	आ	इ	ई	उ	ऊ	ऋ	ॠ	ए	ऐ
a	â	i	î	u	û	ṛi	ṛî	e	ai
ओ	औ	ं	ः	क	ख	ग	घ	ङ	
o	au	ṁ	ḥ	k	kh	g	gh	ṅ	
च	छ	ज	झ	ञ	ट	ठ	ड	ढ	ण
ch	chh	j	jh	ñ	ṭ	ṭh	ḍ	ḍh	ṇ
त	थ	द	ध	न	प	फ	ब	भ	म
t	th	d	dh	n	p	ph	b	bh	m
य	र	ल	व	श	ष	स	ह	ळ	क्ष
y	r	l	v	sh	ṣh	s	h	ḷ	kṣh

ज्ञ

jñ

Lokumanya

Bal Gangadhar Tilak

VEDIC CHRONOLOGY.

बाळ गंगाधर टिळक.

VEDIC CHRONOLOGY

CHAPTER I.

INTRODUCTION.

The ancient Indian literature is divided into two sections–Vedic and post-Vedic–and the chronological sequence of events in the latter can now be pretty accurately determined by referring to the date of Buddha, the invasion of Alexander the Great, the inscriptions of Ashoka, the Shaka era, and a number of other archæological facts recently discovered. But when we go back to the Vedic literature, the oldest portions of which admittedly depict the most ancient Aryan civilization of which any records have been left, we find no such land-marks; and in their absence the method at first adopted by Western Sanskrit Scholars to ascertain the antiquity of the Vedic Civilization was necessarily vague and arbitrary. On the face of it, the Vedic literature is divided into four strata or layers–Chhandas, Mantra, Brâhmaṇa, and Sûtra; and it is evident that each of these stages of development must have lasted, at least for a few centuries before it passed into the next. Taking, therefore, these layers for the basis of his calculation, and assuming, at the lowest, 200 years for each stage of development, Prof. MaxMuller, in his History of Ancient Sanskrit Literature, roughly fixed the age of Vedic civilization at 800 years before Buddha, or at 1200 B. C. The moderation here exhibited, was no doubt unobjectionable, even from a sceptical point of view. But it hardly corresponded with facts; and many other Vedic Scholars even then considered this estimate as too low, and assigning 400 instead of 200 years for each successive stage of development, carried back the antiquity of the Vedic culture to 2400 B. C. This was the general opinion about the antiquity of the Vedic civilization before the publication of my '*Orion*' and Dr. Jacobi's essay "on the age of the Rigveda," in 1893, in both of which the antiquity is carried back to about 4500 B. C. on the strength of astronomical statements contained in the Vedic literature.

Indian astronomy was one of the first subjects which attracted the attention of Western Scholars after the existence of Sanskrit literature became known to them in the last quarter of the 18th century. There are a number of learned and critical articles in the first Volumes of *Asiatic Researches* by Sir W. Jones, Colebrooke, Devies and other scholars on the special and important features of Indian Astronomy, e. g. the lunar zodiac and its antiquity, the Indian and the Arabic divisions of zodiac, the precession of equinoxes, etc, etc.; and so far as we know, there is hardly anything subsequently published, which surpasses these dissertations either in the breadth of view or the soundness of judgment exhibited therein. Even the Vedânga Jyotisha, that small tract on Astronomy appended to the Vedâs, and which is the oldest astronomical work in Sanskrit, did not escape their attention; and, though many verses in it were then obscure, yet it was clearly seen that the position of solstices mentioned therein carried us back, roughly speaking to about 1100 or 1400 B. C. But all the works, examined by these scholars are post-Vedic, and help us little in ascertaining the antiquity of Vedic Civilization. The Vedânga Jyotisha, is certainly the oldest astronomical work now extant; but now that it has been fully deciphered one can easily see that it cannot be the first work of its kind. It must have been preceded by others of its kind; but as these are not now available, we must search the Vedic books themselves for any astronomical data that enable us to ascertain the age of Vedic literature. This was not done till some years later as the attention of early scholars was engrossed in discussing certain side issues which were raised at the time.

The French astronomer Bailey, in his treatise on Indian and Oriental astronomy, published in 1786, had assigned a very high antiquity to the Indian Science; and it was to refute this view that Bentley published his "Historical view of Hindu Astronomy" in 1823. Bentley endeavoured to ascertain the age of the Indian astronomical works solely by comparing the astronomical statements therein with the positions calculated backwards by means of modern astronomical tables and assigning to the said astronomical treatise such time as when the difference between the two was the smallest possible; and by a reckless use of this method he was led to the prepostrous conclusion that many of the ancient astronomical Siddhântâs, were in reality composed or fabricated, as Bentley thinks, in the times of Akabar the great, in order to impose upon the emperor a false notion of their importance and antiquity. He did not stop here, but attacked Sir W. Jones, Colebrooke and other scholars for maintaining the opposite view. This drew forth a sharp reply from Colebrooke who clearly showed how Bentley's method, used exclusively by itself, was utterly unreliable. As Bentley's view is now generally rejected by all scholars, it is unnecessary to go further into the details of the position maintained by him. In spite of its faults, his work however contains some ingenious suggestions which we shall notice later on. It is enough to state here that side by side with this controversy, there was also raised and discussed another important qu estion·viz.-whether the Indian astronomical methods, described in these post-Vedic works were borrowed from the Greeks wholesale or whether the Indian astronomers, who had already a science of their own, improved it by such hints received from Alexandria, as two civilised nations, when they come in contact, are generally glad to receive from each other in the interest of scientific progress. Colebrooke held the latter view; and Rev. Burgen, who translated the

Sûrya Siddhânta while he was a missionary at Ahmednagar in our Presidency, is of the same opinion. But Prof. Whitney, who edited and published the translation with notes under the auspecies of the American Oriental Society in 1860, has not a word to say in favour of the Indian Astronomers, whom he considers incapable of originating any scientific theory or making any even tolerably accurate observations. He has therefore come to the conclusion that the Indians were wholesale borrowers in this respect. The prestige which Whitney enjoyed on account of his great learning and scholarship unfortunately contributed to render, for sometime, his judgment acceptable, in preference to that of Colebrooke. But it has been shown by Shankar Bâlkrishna Diksit, a practical Indian astronomer possessing a wider acquaintance with the whole of the Indian astronomical literature, in his important work on the " History of Indian Astronomy " published in Marâthi in 1896, that Whitney's view is simply the result of his prejudices, that it is entirely opposed to a number of astronomical facts disclosed in the Indian Siddhântâs, and that Colebrooke has said utmost that can be said on the subject. In support of Diksit's rejoinder, we may further mention the fact, overlooked by Whitney and his followers, but noticed by Plunkett in his work on " Ancient Calendars and Constellations, " published in 1903, that in contrast with the praise bestowed by Garga on Yawanas - evidently Greeks - for their proficiency in astrology, we find, the Greek writers of the first century of the Christian era, holding a very high opinion about Indian Astronomy; and that in the life of Apollonius of Tyana, his biographer represents him as learning many things from the Sages of India, especially matters of astronomy. This shows that there was borrowing on both sides and that Whitney's bias against the ancient Indian astronomers was entirely unfounded.

The whole of the above discussion was related and confined only to the post-Vedic astronomical works. But the next question that arose necessitated an inquiry into the astronomical statements contained in the Vedic Works themselves. In 1840 and the following years, the well-known French astronomer, J. B. Biot, published a number of articles in the *Journal des Savants* (subsequently also published in the form of a separate book in 1859), in which he endeavoured to prove that the Indian system of Nakshatrâs must have been borrowed from the Chinese, because in the first place the antiquity of the Chinese system was fully authenticated by reliable ancient texts, going back to 2357 B. C. when the vernal equinox was in *Mao* (i. e. Indian Kṛittikâs), and secondly because from the practical astronomical point of view the stars in the Chinese system of *Sieon* (Nakshatrâs) were mathematically best suited for the purpose for which they were used, viz., to observe meridian passage of equinoctial and solstitial points, as well as that of certain circumpoalr stars. These stars, he maintained, were originally 24 in number to which 4 more corresponding to the two solstitial and two equinoctial points of the time were added, in about 1100 B. C., thus increasing the total number of Sieou to 28. Most of these stars are the same in the Chinese and the Indian system ; but they were unsuited to measure the equal distances between the successive daily positions of the moon, which is the purpose for which they were used in India. Biot, therefore, concluded that the priority of discovery must belong to that place, where the system is found to be best suited to the use made thereof, viz., to China ; and that Indians must have borrowed the same at some later date and used it awkwardly for their purpose. The authority, which naturally belonged to the opinion of so great an astronomer as Biot led some Sanskrit scholars of the time to adopt his view, and we find it accepted, though in a somewhat modified form, even by Whitney in his edition of the translation of Sûrya Siddhânta published in 1860. Prof. Alfred Weber, however, clearly saw the weakness of Biot's position resting as it did, on the supposed antiquity of the Chinese texts ; and the so called convenience of astronomical observation of the time. With great diligence and learning he, therefore, collected all the astronomical statements contained in the various Vedic works, and published in 1860 and 1862 his two essays on "Die Vedisohen Nachricten Von der Naxatra" (the Vedic accounts of the Nakshatrâs). In the first of these he showed how the supposed antiquity of the Chinese texts was unwarranted by historical facts and in the second conclusively proved that the ancient existence of the Indian System of Nakshatrâs, with the Krittikâs at their head, was fully borne out by passages in Vedic works of undoubted antiquity. This was the first time that the astronomical statements contained in the Vedic Works were collected; and so complete is this collection that only a few Vedic texts bearing on the same subject have been since discovered. If Weber had gone further and arranged and co-ordinated the texts collected by him he could have easily perceived that the series with the Krittikâs (Pleiades) at the head was not the oldest of its kind and that the Vedic works expressly refer to a still older system of Nakshatrâs with Mrigshiras (Orion) at the head. But it is not uncommon that a Collector of materials sometimes misses to grasp their true significance, as was the case with the great Danish astronomer, Tycho, whose numerous observations formed the basis of the laws of planetary motion subsequently discovered by Kepler, his successor. Weber had the same low opinion about the capacity of Hindus to make any, even the crudest celestial observations, as was held by Whitney; and though he established the priority of the Indian system of Nakshatrâs over the Chinese, he was, in consequence, led to believe, on almost imaginary grounds, that neither the Indians, nor the Chinese, nor again the Arabs, whose system of Manâzil (Nakshatrâs) resembles the

Indian and the Chinese in many points, were the original discoverers of the system; but that all of them must have borrowed it from some still unknown West Asian, possibly Babylonian source. Prof. Max Muller, in his preface to the fourth volume of the first edition of the Rigveda, published in 1862, contested this view and rejected it as groundless. But he was not qualified to make further research in this matter; and Whitney who was so qualified was prevented by his prejudices from diving deeper into the question, though he clearly saw, as observed by him in his essay on "The Lunar Zodiac" published in 1874 in the second series of his "Oriental and linguistic Studies," that Weber's "theory was no better than a suspicion, perhaps not even worth finding expression as such", or "of a character to compel belief" and that there was no reason "to impugn either the candour or the good sense of any one who might refuse to be won over to a like belief." But if Weber's theory was thus admittedly a mere suspicion it was clearly an error of judgment to refrain, on that account, from critically examining and co-ordinating the Vedic texts, with a view to ascertain which was the oldest system of Nakshatrâs disclosed by them. For the speculative question about the origin of the lunar zodiac cannot be solved satisfactorily without first determining the fact whether the Krittikâ series was the oldest known in India or whether it was preceded by another beginning with Mrigashiras. It is to be regretted, therefore, that Weber's or Whitney's authority diverted, for some time at least, the attention of Western Vedic Scholars from this kind of investigation. But the ultimate discovery of truth is hardly, if ever, prevented by such mishaps. On the contrary we might even say that the path to such discovery often lies though such errors and the progressive elimination thereof. Thus when later researches and discoveries in the Babylonian antiquities failed to bring to light any of those grounds, for the Mesopotamian origin of Nakshatra System, grounds, which Weber and Whitney fondly believed the future would disclose, Thibaut, writing on the subject in the Journal of the Asiatic Society of Bengal in 1894, expressly stated that the theory of the Babylonian origin of the Indian Nakshatra system must, in consequence, be given up; and a year earlier, that is, in 1893 H. Jacobi, who in the meanwhile was prosecuting his investigation into the Rigvedic Calendar, almost simultaneousy but independently came to the conclusion to which I had already arrived, viz ; that in the days of the Rigveda the Vernal equinox was in Mrigashiras or Orion, and that the Vedic texts, properly interpreted, clearly referred to a Nakshatra Series older than the one beginning with the Krittikâs at its head, thereby carrying back the antiquity of the Vedic civilization to the fifth millenium before Christ.

Such is the history of this discovery in the West. In India the course of events which led to it was different. The Vedic texts collected by Weber were not unknown to our old Pandits and the rate at which the equinoctial points retrograded is also so accurately recorded in the ancient astronomical Siddhântâs, that those who would not give any credit for accurate observations to the Hindus, e. g. Whitney, are obliged to account for it, as found by a lucky hit, since even the Greek estimate thereof is far wide of the mark. But it was a fixed article of faith with these Pandits that the Vedas were *anâdi* (beginning-less) being handed over orally from generation to generation, from time immemorial, and in consequence these Vedic texts were never used for any chronological purpose. The introduction of Western education, and with it the Modern historical and critical methods, in our schools and Colleges, have altered this state of things. Those who were educated in these methods, and especially those who had any opportunity to serve in government observatories, were the first to note that Indian almanacs, which, till then, were prepared according to the astronomical tables based on the ancient Siddhântâs and practical works like *Grahalâghava* were faulty and defective, inasmuch as the calculations given therein did not fully correspond to the actual time of occurence of such astronomical events, as the eclipses are the true positions of the planets. And as these almanacs were intended for the timely performance of religious; domestic ceremonies and public festivals it soon became evident to many others that a reform in them was needed. Thus soon after 1850, the late Prof. Chattre, in Poona,

Chintâmani Raghunâthâcharya in Madras and Pandit Bapudevshâstri in Benares came to publish new almanacs mostly based on the British Nautical Almanac. But when such reform was undertaken a controversy soon arose as to whether the tropical or the siderial sphere should be adopted. Indian division of the zodiac into 27 parts, called divisional Nakshatrâs starts from a fixed point in the ecliptic; so that these (divisional) Nakshatrâs represent *fixed* successive positions of the ecliptic each extending over 13°, 20′, and the star after which the division is named, called also the junction star, is situated, not necessarily in the centre, but somewhere between the boundary lines of each division. This is the sphere adopted and described in all the ancient astronomical siddhântâs. Now Indian lunar months are named after the fixed Nakshatrâs at or near which the moon is full during that month; e.g., the month is named Chiatra or Kârtika according as the full moon in the month is found to be near the star Chitrâ or the Kṛittikâs respectively; and the rule is older than Pâṇini. The Indian lunar month thus becomes tied down to a fixed star, or a fixed divisional Nakshatra portion of the ecliptic. But the winter and summer solstitial points as well as the equinoxes on which the seasons depend never remain fixed in the ecliptic. They have a slow backward motion, so slow that it amounts only about fifty seconds a year or one degree in about 72 years. This is known as the precession of the equinoxes, and it causes the seasons to sweep over the whole circle of the ecliptic once in 26,000 years in round numbers. The seasons must, therefore, also sweep through all the Indian lunar months, once in that period. Thus if the lunar months Chaitra and Vaishâkha corresponded with the season of spring (Vasanta) at any time, this season in course of time is bound to fall back in the preceding lunar months. But, if instead of tieing down the divisional Nakshatrâs to a fixed portion of the ecliptic, we count our Nakshatrâs (divisional) from the moveable solstices or equinoxes, the lunar months, named after them, would never cease to correspond with the seasons at first represented by them. By adopting this moveable zodiac the names of the months will have no connection whatever with the fixed stars; in other words, they will be simply arbitrary or conventional. But some astronomical reformers amongst us advocated this course instead of making the seasons shift back through all the months in 26,000 years. On our side Lele, Modak and Shankar Bâlkrishna Diksit were its chief advocates and they termed themselves *Sâyan-Vâdis* or Sâyanists, the word *Sâyana* denoting a preference to the precessional motion (*Ayana*) of the Solstices. On the other the majority of astronimical scholars,—e. g. Chhatre, Bapudevshâstri, and now Pandit Sudhâkar Dwivedi —were of opinion that we should abide by the fixed zcdiac mentioned in the Siddhântas and adopt a siderial sphere, accounting from a fixed point. These are called *Nirayana–Vâdis* or Nirayanists. But if the fixed zodiac is to be retained it was necessary to show how the problem about the shifting of the seasons is to be solved. The motion of equinoxes existed in the ancient times also, and it was naturally supposed that some kind of solution of this difficulty must be found in old works; and if this be found, it will serve as a precedent for us to follow, for great is the value of a precedent in a question which is not purely astronomical but involved important religious issues. The ancient Vedic Books were therefore searched and it was found that corresponding to the different positions of the equinoxes, the Nakshatra Series began with different Nakshatras; and that the months which corresponded with the seasons e.g. Vasanta (spring) were also different in old days. For instance the series of Nakshatrâs began neither with Ashwini nor with Bharani, in old times, but with the Kṛittikâs and previous to it with Mrigashiras. While the Vasanta season is said to comprise either Chaitra and Vaishâkha, or in its stead, Falguna and Chaitra. We shall discuss the full significance of these facts later on. Our object at present is simply to show how the *Sâyana* and *Nirayana* controversy directed the attention of Indian scholars to these facts; and as all engaged in this controversy had a thorough knowledge of astronomy, they did not fail to perceive the importance of these facts from a chronological and antiquarian point of view. But it took some time before the facts were correctly interpreted. Krishna shâstri Godbole, writing on the subject of the antiquity of the Vedas, in the second and third Volumes of the Theosophist, in 1882 (subsequently published as a separate pamphlet) at first believed that when Mârgashirsha was said to be the first month of the year — *Agrahâyani* — it was the first vernal month and consequently referred to the time the autumnal equinox, not the vernal, was in Mrigashiras, thus carrying back the Vedic Antiquity to something like 19,000 B.C. That this was not the right way of interpreting these facts was however at the same time perceived by others. For instance, Narayan Aiyengar and Prof. Rangâchârya, in the Madras Presidency, were of opinion that these texts referred to the time when Mrigashiras (Orion) corresponded with the Vernal and not the autumnal equinox, and the former tried to explain several Vedic myths on this hypothesis, in his "*Essays on Indo-Aryan Mythology*". By this time I too had arrived at the same conclusion by stricter process of reasoning, and in 1892 I sent my essay entitled "*Orion or Researches into the antiquity of the Vedâs*" to the ninth oriental congress held in that year and subsequently published it, with some modifications, in 1893.

Thus it was that Jacobi and myself, working independently arrived at the same result almost at the same time. It was not to be expected that the discovery would pass unchallenged by Vedic scholars, who had hitherto believed that the antiquity of the Vedic civilization could not be proved to be higher than

2400 B. C. But there were some whose literary studies had made them regard this limit as too low; and they welcomed our discovery. Thus Bulher, Barth and Winternitz in Europe and Bloomfield in America expressly declared that they were, in the main, satisfied with the arguments advanced in proof thereof; while Whitney (in the proceedings of the American Oriental Society for March 1894), Oldenberg (in *Zeitschrift* D. M. G. Bd. 49), and Thebaut writing in the Indian Antiquary (Vol. XXIV pp. 85 f) attempted to show that the theory was untenable and unsound. From time to time Jacobi replied to these adverse criticisms. But these replies as well as the other writings on the subject as scattered over different periodicals are not easily accessible to all. I have, in the following pages, therefore, endeavoured to sum up the whole controversy, thus recasting, revising, enlarging, and bringing upto date, the matter contained in my *Orion*. Our conclusions have already been generally accepted by Indian Scholars and the theory is fairly gaining ground in Europe as will be seen from Prof. Louis de la Valle'c Poussin's recent book *Le Vedisme.* The present attempt, will, it is hoped, make it still more acceptable inasmuch as the proofs thereof will be now exhibited in a clear light.

VEDIC CHRONOLOGY.

CHAPTER II

THE VEDIC CALENDAR.

As the method we propose to follow in this book is closely connected with the ancient Calendar, we give here in brief, the leading features thereof. Man naturally measures time by means of the yearly, monthly and daily movements of the Sun, the Moon-with her varying phases-and the fixed stars in the blue vault over his head. Modern astronomy has now most accurately ascertained the periods of these movements. We know, for instance, that the interval between two successive new moons (or two successive full moons) is 29 days, 12 hours 44 minutes and 2.87 seconds. This is called the *Synodical lunar month* and is the one generally used in practice. There is, however, another revolution of the moon called the *Siderial lunar month*, as it measures the time which elapses between two successive arrivals of the moon at the same fixed star, and this is now known to be equal to 27 days, 7 hours, 43 minutes and 11.5 seconds. In other words the Synodical lunar month is longer than the Siderial by 2 days, 5 hours, 0 minute, and 51·37 seconds. Like the moon the yearly revolution of the sun is also two-fold. The time which passes between its two successive arrival at the same fixed star is equal to 365 days, 6 hours, 9 minutes, and 9·6 seconds and is called the Siderial Solar year. But owing to a slow retrograde motion, known as Precession of the equinoxes — which is described in the next Chapter —the Sun arrives at the equinoctial or the solstitial point every year sooner by 20 minutes and 19.9 seconds than the above period. This is termed the *Solstitial* or the *Tropical Solar Year.* It consists, in consequence, only of 365 days, 5 hours, 48 minutes, 49.7 seconds. The difference appears, at first sight, to be very small. But 20 and odd minutes, accumulated for centuries, produce, as will be seen later, serious disturbances in the reckoning of time. Another motion of the Sun, which is of still greater importance to us, is his regular rising and setting, the interval between which we call a *Civil Day.* It is, however, found that all days are not of equal length, being longer in summer and shorter in winter. By the term "day" an astronomer therefore understands the average length of the days during a year and divides this average length in 24 equal hours. The periods of the revolutions of the sun and the moon given above are according to this average standard.

When the periods of the solar and of the lunar revolutions thus became definitely known, it was not difficult to perceive that they were incommensurable between themselves; that is, none of them was contained in the other an exact number of times. It is, therefore, impossible to frame a Calendar that will, without any the smallest correction, hold good for all times. The present Christian Calendar attempts to do this by arbitrarily settling its days and months in total disregard of the moon, and keeping as close as possible only to the solstitial solar year by inserting leap days at certain definite intervals. But even this arrangement is not perfect, since the year, so regulated, is bound to deviate from the true tropical year by about one day in every three thousand years. The correction thus needed is no doubt very small. But it is obtained by sacrificing the moon and a Calendar which dispensed with the moon altogether is utterly useless for regulation of religious sacrifices, ceremonies and festivals most of which depend on the position of the moon in the heavens. The Calendar which was adopted in India in ancient times, and which, with some modifications and additions, is still in force is, therefore, luni-Solar in character, the days and the months being determined by the moon and the year by the sun. It was observed that the moon took about 29½ days to complete her synodical revolution, while the sun returned to the same fixed star in about 366 days. But though the period of 29½ days thus constituted the natural measure of a month, in practice it was necessary to make the month contain an integral number of days; and this number being fixed at 30, it was necessary to adjust this month of 30 civil days with the lunar month of 29½ days. This could only be done by omitting a day in every two months, and the question naturally arose as to what that day should be, so that the two halves of each month might terminate, as closely as possible, with the full and the new moon in that month. Nor was this the only adjustment needed. The yearly rotation of seasons depends entirely on the position of the sun in the heaven, for instance the rains commence and cease, only at a definite period of time in the solar year. And as religious sacrifices were required to be performed also at the proper season, it was necessary to secure the correspondence of the lunar months with the solar year. A year was divided into twelve months. But 12 lunar months (12 × 29½) made 354 days, while 12 Civil months (12 × 30) amounted to 360 days only. Both of them thus fell short of the Solar year of 366 days, the former by 12 and the latter 6 days in a year; and this difference had to be adjusted by inserting additional (intercalary) days or months at the proper time. But a Calendar so regulated, cannot be expected to be permanent. In the first place the periods of the solar and the lunar revolution, on which it was based, were only rough approximations, the solar year being too long by 17 hours, 50 minutes and 50.4 seconds and the lunar month being too short by 44 minutes, 2·87 seconds according to the modern estimates given above. From time to time further corrections were, therefore, necessary, and in the case of the moon this time could not have been very long as the recurrence of the full and the new moon, which could never be mistaken, served as a proper check for the purpose. But not so with the sun. It is true that the seasons are regulated by the position of the sun in his yearly course. But this position does not recur after an absolutely fixed number of days every year. It has a slow backward motion causing, thereby, a retrogression of the seasons which though insignificant each year, amounts to several days in the course of centuries, and it was impossible not to notice the fact of this disturbance, when it amounted to several days, though the cause of it might remain undiscovered. Another correction in the calendar was thus needed, in course of time, and the history of these corrections gives us the data necessary for ascertaining the antiquity of the Vedic civilization. This history cannot be ascertained from the modern Indian astronomical Siddhântâs, which are not only silent on this subject but introduce many new features in the Calendar. For our purpose we must turn to works which are admittedly older than these Siddhântâs, and these works we shall examine here in brief, referring the reader, for further details to the first part of S. B. Diksit's excellent book on the "*History of Indian Astronomy*" previously referred to.

The oldest work on the Vedic Calendar, that we now possess, is a small tract, called the "*Vedânga Jyotisha*" in two recensions, one belonging to the Rig.-and the other to the Yajur-Veda. Its preservation is due to the fact that it is included amongst the six appendices to the Veda (*Vedângas*), which along with the Vedic works are learnt by heart by our priests. The tract was known to Sir William Jones, Colebrooke, Davis, Bentley, and other Scholars of the time and its text was published so early as 1834 by Captain Jervis at the end of his "*Indian Metrology*". But neither these scholars nor Weber, who published the text again in 1862, was able to interpret more than a few verses therein and in consequence the important astronomical statements embodied in these verses did not receive the recognition they deserved. Whitney, for instance, considered that the tract was "filled with unintelligible rubbish" and "left us quite in lurch as regards the valuable information" about the nature of adjustments resorted to in the Vedic Calendar to make the lunar and the solar year correspond with each other. But thanks to the labours of more recent workers in the field—especially of S. B. Diksit, Barhaspatya (Lala Chottelal) and Pandit Sudhâkar Dwivedi,–the difficulties of interpretation have been well nigh completely overcome, and we are now in a position to thoroughly grasp the scheme of the Vedic Calendar set forth in this ancient book. In fact Diksit has already given an outline of the scheme in his above mentioned work on the History of Indian Astronomy. It will be seen therefrom that the Vedânga Jyotisha starts with the data that in a period of five solar years, that is, in 366×5 or 1830 civil days the sun makes five complete revolutions of the Zodiac and that in the same time there are 62 lunar (Synodical) and 67 lunar (Siderial) months. This makes the duration of the Synodical lunar month equal to $29\frac{1}{2}$ days and that of the Siderial lunar month equal to $27\frac{21}{67}$ days. The quinquennial concurrence of the lunar and the solar time here assumed is only approximately true, not rigidly correct; and the daily motions of the sun and the moon, on which the calculations are based, are all mean or average motions, being regarded as constant and not, as is actually the case, varying each day. But this has enabled the 'Vedânga Jyotisha' to frame such rough and ready practical rules, as could be easily worked out by a Vedic priest, and having a knowledge of elementary arithmetic but unacquainted with astronomy, for determining the requisite age (tithi) of the moon and the place (Nakshatra) both of the moon and the sun on a particular day in the period (Yuga) of five years. We might even say that this is the main object of the book and that it does not pretend to be a treatise on astronomy as it was then known. It introduces at regular intervals, two intercalary lunar months, one in the midst and one at the end of a *Yuga* of five years and ordains that every 62nd lunar day (tithi) should be omitted in order to make the different reckonings of time—*Sâvana* (Sacrificial or civil), *Chândra* (lunar), *Nâksatra* (Siderial) and *Soura* (Solar)—correspond with one another. It is expressly stated that the winter solstice in those days occurred when the sun, the moon and the asterism of Dhanisthâ were together; and this Nakshatra is therefore taken as the first point of the celestial sphere which is evidently siderial and not, as supposed by Bentley, tropical in character. Starting from this first point the zodiac is divided into 27 equal parts each named after the principal Nakshatra conatined therein. But to exactly state the mean positions of the sun and the moon in these divisional Nakshatrâs it was necessary to sub-divide; and consequently we find each of the above divisional Nakshatra further divided into 124 parts to define the position of the moon on each day of the Yuga of 5 years. The author of the 'Vedânga Jyotisha,' whosoever he may be, had no knowledge of the gradual retrograde motion of the solstices or the equinoxes. He, therefore, takes the seasons as fixed and gives rules for finding out when each of them commenced and ended. Thus winter or the *Shishira* season commenced with the winter solstice in Dhanishtâ and was followed by *Vasanta* (spring) and *Grishma* (summer); the last ending with the sun at the summer solstice; while the other three seasons, viz. *Varshâ* (rains), *Sharad* (autumn) and *Hemant* (cold) were comprised in the southern passage of the sun, that is, from the summer to the winter solstice. Considerable skill and ingenuity is shown in reducing the arithmetical work to a minimum in order to make the rules as simple as possible, and especially in devising such a conventional serial arrangement of the 27 Nakshatrâs that the numerical order of the Nakshatra in this series may indicate the exact yearly position of the sun in *Amshas* (sub-divisions) of that Nakshatra at the end of each lunar fortnight. But we need not go into these details. Suffice it to say that the word *râshi*, which occurs in this book and which was formerly believed by some scholars to denote a zodiacal sign and betraying thereby a foreign influence, is now found to be used not in that sense at all, but only in the sense of "number" or "quantity" in general; and consequently all conjectures about the date of the book based upon this misconception must be set aside as groundless.

Such, in brief, is the scheme of the Vedic Calendar disclosed by the oldest tract on the subject. But as the astronomical elements on which it was built up were only approximately true and further approximations were accepted to simplify the rules, it was not to be expected that the scheme would remain in force for more than two or three generations without any further modifications or corrections; and as pointed out by S. B. Diksit, we do find such modifications introduced in later times. For instance, we find a new 12 years' cycle of Jupiter introduced at about the same time, doubtlessly to secure a still greater coincidence of the lunar with the solar time than a five years cycle would do; and in the Pitâmaha Siddhânta, as epitomised by Varâha Mihira in his Pancha-Siddhântika, rules for omitting a tithi (tithi-Kshaya) or introducing an intercalary month are given in a slightly modified and more correct manner. But we are not here concerned with these later developments of the Vedic Calendar. We have to see what kind of Calendar was used in the time of the Brâhmaṇâs and the Vedic Sanhitâs; and for this purpose we may well make the "*Vedânga*

Jyotisha" as our starting point and see what elements of the system contained therein are found in the older Vedic Works.

This work has been done by S. B. Diksit in the first part of his work on the history of Indian astronomy, referred to above. He has shown that in the Brâhmaṇa period of Vedic literature the nature of the Calendar was substantially the same, as we find it in the "*Vedânga Jyotisha*". Thus the list of Nakshatrâs given in the Taittiriya Sanhitâ (4. 4. 10), Taittiriya Brâhmaṇa (1. 5. 1.; 3. 1. 1–2.; and 3. 4. 4–5.) and even in the Atharva Veda Sanhitâ (19.7) is the same as in the 'Vedânga Jyotisha,' with this difference, however, that whereas Dhanisṭhâ is the first in the 'Vedânga Jyotisha', the series in the Brâhmaṇa and the Sanhitâs always begins with the Kṛittikâs. (INCOMPLETE.)

SYNOPSIS
OF
VEDIC CHRONOLOGY

(PRELIMINARY SKETCH.)

CHAPTERS

(I to IX)

with Appendixes.

VEDIC CHRONOLOGY.

CHAPTER I.

INTRODUCTION.

1 Importance of the subject.

2 Hindu view — Eternal.

Shastric view — Yugarambha.

3 Literary European view (Max Muller, Haug, Whitney ,Weber).

4 Astronomical —

(*a*) Colebrooke, Bentley, Hastings, Deivis, Jones.

(*b*) Biot (Journale des Savounts).

(*c*) Weber, Whitney, Max Muller.

(*d*) Jacobi, Tilak, Oldenburg, Thebaut (Grandness), Whitney, Barth, Bulher (Indian Antiquary), Bloomfield.

(*e*) Indian Astronomers — Bapudev Shastri, Chhatre, Raghnath, Modak, Lele, Ketkar, S. B. Diksit, Rangacharya, Narayan Iyengar, Tilak, Bombay Calendar Conference, Pandit Sudhakar, Krishna Shastri Godbole etc.

[Captain Jervis published Lagodha Jyotisha in his "Indian Metrology" in 1834.

Asiatic Researches started 1793 (vol. of 1798 astronomy hints only-page 404.)

Bailey; 1787—"Taite'de l'astronomie Indieane et Orientale."

Colebrooke; Davies. Bentley, 1823. Colebrooke's reply to Bentley, 1826.

J. B. Biot 1840—(1859). Weber 1860-61. Burgess—Whitney 1860. Max Muller 1862. Whitney Lunar Zodiac 1874. S. B. Diksit 1896. Jacobi and Tilak 1893.]

CHAPTER II

MEASUREMENT OF TIME

(VEDIC CALENDAR)

(ASTRONOMICAL)

General astronomical discussion regarding the measurement of time. Day, night, month, seasons, year, eras, cycles, Kali, Precession of the equinoxes, Motions of the Earth—yearly and daily, the Sun, the Moon etc. Vedic Calendar published by Captain Jervis at the end of his "Indian Metrology" 1834. Weber 1862, Thibaut 1877, Godbole 1882, Diksit 1896, Barhaspatya 1906, Sudhakar 1907.

Detailed explanation of the consequences of the precession of equinoxes and the Nakshatra system. Equating the lunar and the solar year:—

	D.	H.	M.	S.
Lunar Month (Siderial) =	27 —	7 —	43 —	11.5
Lunar Month (Synodical, from New Moon to New Moon) =	29 —	12 —	44 —	2.87
= 29·530589 days.				

	D.	H.	M.	S.
Solar year (Siderial =	365 —	6 —	9 —	9·6
„ „ (Tropical =	365 —	5 —	48 —	49·7
The difference =	0 —	0 —	20 —	19.9

	D.	H.	M.	S.
The annual retreat of the equinoxes =				-50.10
Time taken by the sun to describe this arc of the ecliptic=			20 —	19'9
Solar Siderial year =	365 —	6 —	9 —	9'6
Lunar Synodical year	354 —	8 —	48 —	34'44
The difference =	10 —	21 —	20 —	35·16

TABLE I.

Name of the Book.	Com. of Uttarâyaṇa (Winter Solstice.)	Nakṣhatra of Winter Solstice.	Com. of Dakṣhiṇâyana (Summer Solstice)	Nakṣhatra of Summer Solstice.	Remarks
1 Varâhamihira.		*उत्तराषाढा (अर्ध)		पुनर्वसू	(Mentions older posi) V. E. at the end of रेवति.
2 Vedâṅga Jyotiṣha Verse 5.	माघ शुक्ल १. (अमांत मास)	श्रविष्ठारंभ or धनिष्ठा.		आश्लेषार्ध	V. E. in 10 भरणी
3 Shankhyana Brâhm. 19.2, 3 and श्रौत 13.19.1	माघ व. ३० (पौर्णिमांत मास) that is माघ शु.	,,	,,	,,	See Weber p. 345 note 4. & p. 353.
4 Garga	माघ शुक्ल १	श्रविष्ठारंभ		,,	Orion p. 36. Weber p. 355; Note 1.
5 Parâshara	,,	,,		,,	Weber p. 355.
6 Mahâ-Bhârata Anu. P. 167.26	माघ शुक्ल ७		,,	मघा Anu. P. 88.12	Series begins with कृत्तिका Anu. p. 89.
7 बौधायन श्रौत					See Aiyer's V. Era
8 मैत्र्युपनिषद् 6.14	माघ शु. १५	धनिष्ठार्ध		मघारंभ	
Mahâ. Bhâ.		*श्रवण			Ashv. P. 44.2 Adi. P. 70.34.

CHAPTER III

THE YEAR BEGINNINGS.

1 According to seasons.

2 According to solstices.

3 Devayâna and Pitriyâna — Description of—Shat. Brah.

4 Sacrificial.

CHAPTER IV

VEDANGA JYOTISHA.

1 Equinox in Revati —

Intermediate stage in M.B.

2 Winter Solstice in श्रवण.

3 Winter Solstice in धनिष्ठा.

Authorities—Vedânga Jyotisha, Parashara, Vasistha, Garga, Mahabharata, Baudhayana, Varaha Mihira, Sushruta, Maitrayani Upanishad (older).

(VIDE: TABLE No. 1)

CHAPTER V

THE KRITTIKAS

1 Krittikas head the Series मुखऋतूनाम्.

2 Devanakshatras begin with Krittikas.

3 Passages from Tait. Sanhita, and Tandya Brahmana.

4 Krittikas in the East. Shat. Brah., Kathaka and Shalvasutras.

5 माघी पौ. (The beginning of the year) सांख्या. श्रौत (15-12-17), Weber p. 335. Bulher p. 243 quotes Vasistha XI. 40.

6 मघा नक्षत्र. Dakshinayana. M.B. अयनं मघा मे. Atha. Veda.

7 विशाखा–वैचित्रौ. Other nations also reckoned with Kritt. in the east. Alberum Hermes African note. Chinese.

8 Position of climes in Maitra. Upanishad.

मघासु सर्पिसंयुक्तं पायसं दक्षिणायने (दानं कुर्यात्)

अयनं मघा मे Atharva Veda; M.B. अनु. ८८.१२.

Vasistha 11·40.

माघी च पौर्णमासीयं &c. in M.B. अश्वमेध ८५.८, p.167.

CHAPTER VI

MRIGASHIRAS.

(1) आग्रहायणी, (2) फाल्गुनी पौर्णिमासी, (3) भाद्रपद, rainy season —श्राद्धकाल Manu 3.259, Vassa of Jains and श्रावणी Per. आश्वथिक Ramayan, (3) Parsi Calender, (5) Mula, Jyestha. (Weber's error).

Prajapati running over to Rohini.

Other confirmatory evidence :—

1 Chaturmasya. 2 Dhruva Polar star. 3 तिष्यगुरुसंयोग –Ketkar. 4 Names of planets मघाभू etc. Bentley. 5 आश्वत्थिक month. 6 Ram. Nakshatras. 7 रोहिण शंकाभेद 8 Chaldean Inscriptions.

(12) पुछं पूर्वे. *Ram. Nakshatras.*

Gopath Br. (I. 19) मुखमुत्तरे पुछं पूर्वे

Shankhyayan (5·1) „ „

Taitt. Br. 1.1.2.8 „ „

Shata patha 6.2.2.18 „ „

Simply फाल्गुनी पौर्णिमासी मुखं with distinguishing between उत्तरे and पूर्वे

Taitt. Sanhi. VII. 4. 8 मुखसंवत्सरस्य.

Tandya Br. V. 9.

Shat P. Br. VI. 2. 2. 18

Shankhy. Br. IV. 4.

Kathak. Sam. 8. 1 ऋतूनामुखं.

CHAPTER VII

VEDIC TEXTS

1 Marriage of Surya Sâvitri (द्वादशस ऋतुं)

2 पर्जन्यवृद्धं महिषं

3 श्वानवस्तो बोधयितारं

4 Vrishâkapi hymns.

5 Weber's view that N. were unknown in Vedic times. फल्गुनी, अ।श्वाधिक names.

6 Dhruva Star.

अर्जुनस्य इमे बाणाः नेमे बाणा शिखंडिनः ।
कृतंति मम गात्राणि माघमासे गवामिव ॥
भीष्मपर्व १२०.६४

मन्येऽहं पांडवान् सर्वान् भारद्वाज शरार्दितान्।
शिशिरे कंपमाना वै कृशागाव इव प्रभो ॥
द्रोणपर्व १५५.६–७

गृहस्नेहावबद्धानां । नराणामल्पमेधसां ।
कुक्षी खादति मांसानि । माघमासे गवा इव ॥
शां० १३९.८८

तैषी अमा॰ } पाष
तैषी पौर्णि॰ }
पौष { तैषी पौर्णि॰ … माघी अमा॰
माघी अमा॰ } माघ
माघी पौर्णि॰ }
माघ { माघी पौर्णि॰ … फाल्गुनी अमा.
फाल्गुनी अमा. } फा.
फाल्गुनी पौर्णि. }

and so on.

CHAPTER VIII.

TRADITIONS AND MYTHOLOGY.

1 Orion Tradition.

1 Prajapati and his daughter.

1 Vritra and Namuchi.

1 The Moon and Rohini.

1 Krittikas M. B.

CHAPTER IX.

1 Continuous change and withdrawal; precession of seasons.

2 The period beyond all this: A Aditi period.

3 Literary evidence, Bulher, Inscriptions—especially Chaldean.

4 Different periods.

5 Conclusion.

APPENDICES.

I Chaldea and India.

II (*A note on* मघासु गावो हन्यंते.)

माघमासे गवामिव Mahâ Bhârata (भीष्मपर्व Ad. 120. V.64 p. 356; and शांतिपर्व Ad 139. V. 88, p. 333):—

अर्जुनस्य इमे बाणाः नेमे बाणाः शिखंडिनः ।
कृतंति मम गात्राणि माघमासे गवामिव ॥

(भीष्म-पर्व)

गृहस्नेहावबद्धानां नराणामल्पमेधसां ।
कुस्त्री खादति मांसानि माघमासे गवा इव ॥

(शांति-पर्व)

This shows that in मघासु गावो हन्यंते (Rig-Veda) we must understand मघा to mean माघी पौर्णिमा or *lunar* postition in मघा and not *solar*. The same conclusion follows *firstly* because marriages are not celebrated when the *sun* is in मघा, for then the moon to be full must be in श्रवण and so the month would be श्रावण, and no marriages are celebrated in श्रावण. *Secondly* फाल्गुन is a month of marriage ; but if we interpret फल्गुनीषु च उह्यते to mean a solar नक्षत्र the moon to be full must be in भाद्रपदा and the month will be भाद्रपद which again is not a marriage month. *Thirdly*, as it is a marriage between सोम (masculine) and सूर्या (feminine) we must take it to refer to the *full moon night*; when the moon is the brightest and the sun invisible—hence सूर्या as feminine.

If the passage be so interpreted, as it must be, for reasons given above, then it confirms the statement फाल्गुनी पौर्णिमासी संवत्सरस्य प्रथमा रात्रिः in the Brahmanas.

Two interpretations possible :—(1) सूर्य-नक्षत्र or (2) चंद्र-नक्षत्र.

I सूर्य-नक्षत्र—so that Surya is in मघा and फल्गुनी successively.

Pros.—(*a*) If we take भाद्रपद कृष्ण ३०, then the sun and the moon are together.

(*b*) The fortnight preceding is *Pitri-paksha* when the sun is at the end of Maghâ or the beginning of Purvâ फल्गुनी.

(*c*) The condition that फल्गुनी is a विवाहनक्षत्र is satisfied.

(*d*) द्वादशस्य ऋतुं in मंडूक सूक्त corroborate.

(*e*) पर्जन्यवृद्धं महिषं is satisfied (Rig. 9.113.3.)

(*f*) फल्गु = नक्षत्र because figs were ripe when सूर्य in फल्गुनी.

(*g*) Sâvitri व्रत commences in Rainy season.

II. In the case of चंद्र-नक्षत्र we take (*a*) माघी पौर्णिमा and (*b*) फाल्गुनी–पौर्णिमा.

Pros:—(*a*) फाल्गुनी पौर्णिमासी संवत्सरस्य प्रथमा रात्रिः ।

(*b*) मघासु गावो हन्यंते । फल्गुनीषु च उह्यते । माघमासे गवामिव । M.B.

(*c*) फल्गुनी is a विवाहनक्षत्र (lunar.)

(*d*) As it is a marriage of चंद्र, the moon must be a full moon and Surya dark or invisible, shrouded or veiled.

Cons :—(*a*) Feast is not properly explained. माघ is not a festive month.

(*b*) the Sun and the Moon are not together on a पौर्णिमा day.

(INCOMPLETE.)

Points of Difference between Greek and Hindu Astronomy.

1 The System itself—Greek partly heliocentric, Hindu entirely geocentric.

2 Names of Planets different in each.

3 Periodical revolutions and distances of the planets different.

4 Hindus use signs, Greek chords.

5 Hindu calculating method algebraical, Greek geometrical.

6 शीघ्र व मंदकर्म Epicycles, motions of apsides.

7 Week days.

8 Zodiacal Signs. Naksatrâs

9 Vikshepakas = Inclination to the ecliptic.

Similarities between Vedanga Jyotisha and Surya Siddhanta.

1 Yuga:— At beginning सूर्य and चंद्र are together in Vedang Jyotisha and all planets in सूर्यसिद्धान्त.

2 The four time-measures सावन, सौर, चांद्र and नाक्षत्र.

3 Calculations for the whole *yuga*.

4 Arithmetical treatment.

NOTE

ON THE INTERPRETATION OF

THE

VEDANGA JYOTISHA.

(CRITICISM AND SUGGESTIONS.)

By **B. G. TILAK.**

MANDALAY;

15-5-14.

NOTE
ON THE INTERPRETATION OF
THE VEDANGA JYOTISHA.
(CRITICISM AND SUGGESTIONS.)

Small as it is, the Vedanga Jyotisha is the oldest tract on Hindu Astronomy, and its importance, from a historical point of view, especially of the position of solstices recorded therein, did not escape the attention of early Sanskrit scholars, like Colebrooke, Sir William Jones and others. The tract appears in two recensions, —the Rik. or the one belonging to the Rig-veda, and the other to the Yajurveda. Of these the first was published by Captain Jervis, at the end of his *Indian Metrology*, so far back as 1834, and the late Prof. Weber brought out in 1872 a critical edition of both the recensions, with various readings collected from the manuscripts then available to him. But the corrupt state of the text, as well as the enigmatical nature of the rules contained therein, made the work —except a few simple verses — quite unintelligible; and some scholars even doubted the antiquity of the astronomical statements embodied in it. Dr. Thibaut, in his essay on the Vedanga, which appeared in the Journal of the Asiatic Society of Bengal in 1877, was the first to decipher a few of the difficult verses, and among others who followed him, the name of the late Shankar

Balkrishna Dıksit deserves to be specially mentioned. Mr. Diksit, in his important work on *The History of Indian Astronomy*, written in Marâthi and published at Poona in 1896, has not only explained more verses of the Vedanga than any of his predecessors, but has given us for the first time a lucid account of the principles and the methods of calculation adopted in the Vedanga, together with an outline sketch of the quinquennial calendar based upon it. Whitney's reproach that the Vedanga was "filled with unintelligible rubbish," was thus proved to be entirely groundless. But still about 12 verses remained unexplained; and the credit of explaining these for the first time, undoubtedly belongs to Lala Chhote Lal (*nom de plume* Bârhaspatyah), an executive engineer in N. W. P., who published his valuable essays on this subject, first in the *Hindustan Review* in 1906, and subsequently, with certain additions and corrections in a book form in 1907. But in a task beset with so many difficulties, no finality, as Bârhaspatyah himself acknowledges, was to be expected; and we, accordingly, find Mahâmahopâdhyâya Sudhâkar Dwivedi, the well known learned Pandit of Benares, endeavouring to improve upon Bârhaspatyah's defective explanations, in what he calls his own (Sudhâkar) Bhâshya on the Vedanga, published in a pamphlet form, along with Somakara's Bhâshya, in 1908, at the Medical Hall Press, in Benares. Bârhaspatyah, in reply, has attempted to defend his own interpretation of the text, and the controversy has unfortunately assumed a somewhat personal aspect in N. W. P. We are not, however, concerned with this aspect of the question. It is true that Pandit Sudhâkar has not succeeded in giving us a more rational or simple explanation of the Vedânga verses, except mostly by making ingenious but radical changes in the traditional text. But still his work has a value of its own, especially in drawing our attention to the weak points in Bârhaspatya's work; and subsequent critics have to see whether these defects cannot partially, if not wholly, be removed without any violent amendment of the text. This is what the present note attempts to do, especially in the case of the nine verses mentioned in the preface to the Sudhâkar Bhâshya, as those wherein the Pandit considers Bârhaspatya's explanations as seriously defective. Not that there are no more points of difference between them. But these being of minor importance are not discussed in this note. To avoid constant repetition of full names I have, in the sequel, used the letters *B* and *S* to denote Bârhaspatya and Sudhâkar or their works mentioned above, while the Rik and the Yajus recensions of the Vedânga are indicated by the letters *R* and *Y* respectively.

The astronomical elements on which the Vedânga bases its calculations are only approximately true; and we shall see later on how in the case of the moon, at least, a correction was provided for, when the error became too obvious to be neglected. We might, however, generally say that the Vedânga lays down rules to calculate first the fortnightly, and then from it the daily position of the sun and the moon from these approximate mean data. The positions to be thus determined are two-fold. First we determine the position of the sun and the moon, in space, that is, amongst the celestial Nakshatras (*a*) at the end of each parvan and (*b*) tithi. But a parvan or a tithi hardly ends with a Nakshatra. We have, therefore, secondly to ascertain the time when the sun or the moon first enters into (*a*) the parvan-, or (*b*) the tithi-nakshatra. In other words we have to ascertain not only the Nakshatra at the end of a parvan or tithi, but also the time elapsed between the entry of the sun or the moon into that Nakshatra and the end of the required parvan or tithi. The first is done by ascertaining the Nakshatra *amshas* and the second by determining the daily *Kalâs* in the case of the Moon, and in the Sun's case the daily *amshas* or the hour angle *amshas* as *B* calls them. The Vedânga provides rules for all these purposes. Of these the first and the principal one directs us to find the Sun and the Moon's parvan position, or their Nakshatra amshas, at the end of any given parvan, and is contained in the following verses:—

भांशाः स्युरष्टकाः कार्याः पक्षा द्वादशकोद्गताः ।
एकादशगुणस्योनः शुक्लेऽर्धं चैंदवा यदि ॥ R. 10; Y. 15.

In the place of गुणस्योनः *Y* reads °गुणश्चोनः, and adopting this reading *B* has on the whole correctly interpreted the first half of the verse, though his *anvaya* is a little strained; while as regards the second half, his interpretation is both laboured and defective. *S* tries to improve by changing स्योन: into श्रेन्दो:, अष्टका: into इष्टका: and taking द्वादशक to mean a year of twelve months. But with all these changes, the verse does not give us the parvan amshas of both the sun and the moon, but only of the latter, while the reading श्रेन्दो: renders the words इंदवा यदि at the end of the verse entirely superfluous. *S*'s interpretation cannot, therefore, be considered as satisfactory; at any rate it is not an improvement on *B*'s. Both have, in my opinion, unfortunately missed the true meaning of ऊन in this verse, and that is the main cause of the laboured construction adopted by one and the bold emendations proposed by the other. ऊन does not here signify 'minus' as *B* takes it, nor is it necessary to be altered into श्रेन्दो: as *S* has done. ऊन literally means '*less*' or '*remaining*'; and when we take out a number or its complete multiples from another what remains may very well be called ऊन, not absolutely but *relatively* to the first number, because it must always be *less* than the latter. Thus, in the present case, ऊन means the remainder after one or more dozens (द्वादशक) are taken out of a given number of Pakshas; and in this sense it exactly corresponds with the English word 'odd', in the phrase 'a thousand and odd'. The Vedânga divides a given number of Pakshas into so many dozens and so many odd, e.g., $93 = 84 + 9 = 7$ dozens (द्वादशकाः) and 9 odd (ऊनाः); and then calculates Nakshatra amshas first for the dozens and then for the remaining, or the *ûna*-pakshas as it calls them. One operation is thus split into two subordinate parts in order to cut short the labour of arithmetical calculation. For, as the amshas at the end of a dozen pakshas are found to be always equal to 8 or the corresponding multiples of that number, (if we omit the complete Nakshatras, that is, the multiples of 124 amshas from the total), it is obviously easier to get the amshas at the end of any dozen pakshas multiplying the dozens by 8, than by multiplying by 11 the total number represented by the said dozens. But in this case it is necessary to say how the amshas for the pakshas in excess of a dozen, i. e., the *ûna*-pakshas are to be counted; otherwise the rule would remain incomplete, that is, inapplicable to *any* given number of pakshas in general. Taking all these things into consideration and changing पक्षाद्द्वादश° into पक्षद्वादश°, I would, therefore, read the verse as follows :—

भांशाः स्युरष्टकाः कार्याः पक्षद्वादशकोद्गताः ।
एकादशगुणश्चोनः शुक्लेऽर्धे चेंदवा यदि ॥

and without any strained *anvaya* translate :—" the (parvan) Nakshatra amshas should be made by (that is, counted) groups of eight, (each) arising out of (each) dozen (द्वादशक) pakshas, and by (adding to it) the eleven-fold of the excess remaining (ऊन); and again (add) half (the nakshatra–amshas i.e. 62) in the case of the bright (paksha), if the moon's amshas (therein are required)." The first three lines of the verse give a general rule, which is applicable both to the Sun and the Moon, while the last line states, by way of an exception, the change necessary to be made therein if the Moon's amshas, at the end of a bright fortnight, are alone required. For example, suppose, we have to find the Nakshatra-amshas at the end of 93rd parvan. Here 93=7 dozens and 9 odd (*ûna*). Therefore the amshas are equal to $7 \times 8 + 9 \times 11 = 56 + 99 = 155$, or, deducting 124 therefrom, 31 only. These are the parvan-amshas in general. But 93rd is a bright paksha; and if the Moon's amshas alone are required, we must add 62 to the above result; for then the Moon is in opposition to the Sun, that is, 13½ Nakshatras apart, or if we compare the Nakshatra-amshas only, she has 62 more amshas than the Sun. A complete table of the parvan-amshas for all the 124 pakshas of a *Yuga* can thus be very simply constructed by the short and simple rule contained in this single verse. It is unnecessary to give here the reasons on which this elegent rule is based; for they are fully explained both by *B* and *S*, and also by Mr. Diksit before them. There is no difference of opinion as to the main object of the rule. What is the meaning of ऊन, and how to interpret the verse so as to give us a plain rule are the only points in dispute; and the course I have indicated, will, I believe, be found less objectionable than any hitherto proposed. The traditional text is only slightly altered, while the rule is made not only complete but also as comprehensive as possible without straining the ordinary meaning or the construction of the words in the text.

The parvan amshas of the Sun and the Moon being thus determined, the Vedânga makes them the basis of calculations for finding the rest of the lunar and solar positions mentioned above. Thus the next verse in the Rik. recension tells us how to calculate, from the parvan-amshas so ascertained, the time in *kalâs* of the Moon's entry into the last parvan Nakshatras. The method followed is the same as in the last verse. The required time in *kalâs* at the end of a dozen pakshas is first determined and then we are directed to make certain additions thereto in order to find the required *kalâs* at the end of any remaining odd or *ûna*-paksha. The verse is as follows :—

कार्याः भांशाष्टकास्थाने कला एकात्रविंशतिः ।
ऊनस्थाने द्विसप्ततीरुद्वपेद् ऊनसंमिताः ॥ R. 11; Y. 19.

The word भांशाष्टका° clearly refers to the process of calculating the *bhamshas* given in the previous verse; and, thus interpreted, the first half of the verse plainly directs us "to substitute 19 *kalâs* in the place of a group of eight Nakshatra-amshas." The word कलाः further shows that we are here dealing with the case of the Moon; and since, as stated in the previous verse, a group of eight Nakshatra-amshas, is to be taken for each dozen pakshas, the first half of the verse practically lays down the rule that "to find the time (in *kalâs*) of the Moon's entry into the last parvan Nakshatra at the end of a dozen pakshas, one has only to substitute *19 kalâs*, in the place of every group of eight Nakshatra-amshas which correspond to the required dozen pakshas." *B* has rightly interpreted this part of the verse, after making the necessary correction of altering भांशाष्टका° into भांशाष्टक°. He has also shown that the rule is derived from or based on a sound mathematical reasoning. At the rate of 1809 Nakshatras in 124 pakshas of a Yuga the Moon traverses $175\frac{8}{124}$ Nakshatras in a dozen paksha; and since she takes one day and seven *kalâs* to move through a Nakshatra (R. 18; Y. 39), it will be found that to complete 175 Nakshatras she requires 177 days and 19 *kalâs* of the 178th day. In other words the time of the Moon's entry into the 176th or the last parvan Nakshatra at the end of a dozen pakshas is 19 *kalâs* of the corresponding parvan day; and similarly if the number of pakshas be two dozen, the number of *kalâs* would be 2×19 or 38, and so on. But having once missed the true meaning of ऊन *B* has unfortunately failed to follow up this course in interpreting the second half of the present verse; and *S* is led to radically alter the whole verse to suit his guess about its contents. He cannot make anything out of the verse except by artificially interpreting अष्टका to mean 4, and changing ऊनसंमिता: into इन-

संमिता :, the word इन in the latter case being taken to mean the number 12. When one must resort to such artifices to make a verse intelligible, one may, I think, very well suspect that he has missed its right meaning. The rule obtained by *S* after so many changes, does not again give us the time of the Moon's entry into the last parvan Nakshatra; but only enables us to convert the Nakshatra-amshas into *kalás* thereof, which is quite besides the purpose in view. In my opinion the second half of the verse is supplementary to the first; and that after stating the rule for converting *bhamshas* into *kalás* at the end of a dozen pakshas, the Vedânga proceeds to deal with the case of pakshas in excess thereof—the ऊन pakshas as it calls them. How the second half of the verse should be interpreted according to this view, will be seen from what follows:—

In the Rik. recension this second half of the verse is thus given:—

ऊनस्थाने द्विसप्ततीरुद्वपेद् ऊनसंमिताः ।

and I think it is the correct reading. The Yajus text has द्विसप्ततिमुद्व॰ and, adopting it, both *B* and *S* consider द्विसप्तति as a Numeral adjective qualifying कलाः or काष्ठाः understood, though eventually they differ widely in their interpretations. At the first sight this seems to be the proper construction of the verse; for, according to Sanskrit grammar, numerals like द्विसप्तति are declined in the singular number, even when they are in apposition to a plural noun. The same grammar, however, teaches us that these numerals are used also to denote independent numbers, and that the plural form is not only correct, but necessary when the speaker wishes to indicate "many or several groups" thereof. For instance, if we want to denote "several groups of twenty", विंशतय: is the proper form to be used. Without supplying कला: or काष्ठा: as understood, I would, therefore, take त्रिसप्तती: as an independent numeral in accusative plural, governed by the verb उद्धरेत् and qualified by the adjective ऊनसंमिता: used here not in the nominative, but in the accusative case. This construction is at once simple and natural. It is necessary further to note that in contrast with °स्थाने कार्या: in the first half of the verse, we have, in the second °स्थाने उद्धरेत् evidently meaning that here द्विसप्तति's are not to be *substituted for* (कार्या:), but *added to* something else. Thus the line means that "one should raise up, that is, increase (the figure) in the place of ऊन by (as many) seventy-two's (as are) equal in measure to the ऊन (number)." For example, suppose we wish to ascertain the time (in *Kalás*) of the Moon's entry into the last Nakshatra at the end of 18 pakshas. Here 18 is equal to one dozen *plus* six (ऊन) pakshas. The *Kalás* at the end of one dozen pakshas, are equal to 19 according to first half of the verse. What remains is to find out the number of *kalás*, required to be added to the above result on account of the six additional (*ûna*) pakshas. For this purpose the Vedânga now directs us to raise up the ऊन Number, here equal to 6, by as many 72's as are equal to the ऊन Number itself, that is by 6×72 in the present case. We thus get $6+6\times72$. But $6+6 \times 72=6 (1 + 72) = 6 \times 73$; or generally $n + 72n = 73n$. Therefore the rule practically comes to mean that the *ûna* number should be multiplied by 73. The result, be it remembered, does not give us the number of the required *Kalâs*. It is only the first, or the preliminary step in the calculation required to be made for that purpose. The Moon traverses 1809 Nakshatras in 124 pakshas, that is, $14\frac{73}{124}$ Nakshatras during one paksha. Or one may say that over and above 14 complete Nakshatras she traverses a fraction of a Nakshatra equal to 73 amshas per paksha. Therefore, in a given number of *ûna* pakshas, this fractional part of a Nakshatra or the भशेष of the Moon, as it may be fitly termed, would, in amshas, be equal to 73 multiplied by the said number of the pakshas. This is exactly the rule contained in the half verse under consideration. The next step is to derive the Moon's required *kalâs* from this result. But before we pass on to it, it is necessary to examine a little more closely, the nature of the भशेष thus ascertained.

The moon passes from one Nakshatra into another in succession. Therefore the time of her entry into a particular Nakshatra is the same as the time of her arrival at the end of the previous Nakshatra. Thus, in a single paksha the Moon traverses 14 complete Nakshatras and a fraction of the 15th, equal to 73 amshas; and so the time of her entry into the last or in this case the 15th Nakshatra, is the same as that of her arrival at the end of the previous 14 Nakshatras. But though the fractional part in excess of 14 Nakshatras may thus be neglected in this case, yet in the case of the second and the subsequent pakshas it must be taken into account, for the accumulated fraction then gives rise to additional complete Nakshatras. For example, the Moon traverses 2 ($14\frac{73}{124}$) or 28 Nakshatras and 146 amshas in two pakshas. Here 146 amshas are equal to 1 Nakshatra and 22 amshas. Therefore the Nakshatras completely traversed by the Moon during two pakshas is not 2 × 14 or 28, but 29. So the time of her entry into the last Nakshatra at the end of the second paksha is equal to the time of her arrival at the end of 29th and not the 28th Nakshatra; and so on in the case of succeeding pakshas. In short, we must multiply 73 by the number of pakshas, see how many complete Nakshatras are contained therein and add the latter to the corresponding multiple of 14 to find out the complete Nakshatras traversed by the Moon at the end of a given number of pakshas. This is the main object for which the भशेष is calculated. Or we may obtain the same result in a different way. 73 is equal to 62 + 11; and 62 amshas are equal to half a Nakshatra. Therefore we may say that in addition to

14 Nakshatras the Moon traverses ½ Nakshatra + 11 amshas in the first paksha. For the second paksha this *excess* will be doubled, that is, it will be equal to 1 Nakshatra + 22 amshas; for the third tripled or 1½ Nakshatra + 33 amshas and so on. These results are embodied in the following table*. As the number of *ūna*-pakshas is always less than a dozen, the table need not be calculated for more than 11 pakshas.

The figures in first part of column 3, that is in column 3 (a), show the number of additional Nakshatras, arising out of the अंशेष in each case. Thus for the first paksha the अंशेष does not amount to a complete Nakshatra; for the second it gives rise to one complete Nakshatra; for the fifth (omitting one half) to 2 complete Nakshatras and so on. Or, generally speaking, we may say that the number of complete Nakshatras arising out of the अंशेष is equal to the integral part of half the number of pakshas; e.g. for the fifth pakshas, it is equal to 2, the integral part of 2½; and so on. There is, however, an important exception to this rule. If the amshas given in column 3 (b) exceed 62 or half a Nakshatra, these amshas, when added to half the Nakshatra in column 3 (a) may increase the number of the complete Nakshatras. For example take the 7th paksha. Here half of 7 is 3½ and this is the figure entered in column 3(a) opposite to the 7th paksha. But the number of amshas given in column 3 (b) is 77, or half a Nakshatra plus 15 amshas. There-

* Vide Table II p. 59.

TABLE II

No. of ऊन pakshas.	The corresponding भशेष in multiples of 73.	The same भशेष in Nakshatras and amshas.		Remarks.
1	2	3		4
		a	b	
1	$1 \times 73 = 73$	$\frac{1}{2}$	11	*Note* that the amshas in column 3 (b) correspond to the *ūna*-pakshas only. If the total amshas at the end of a given number of pakshas be required, we must add to these the amshas at the end of the previous dozen pakshas. Thus the amshas at the end of 18 pakshas (or 1 dozen plus 6 pakshas) will be equal to 8 amshas for the dozen plus 66 amshas for the 6 *ūna*-pakshas; and so on.
2	$2 \times 73 = 146$	1	22	
3	$3 \times 73 = 219$	$1\frac{1}{2}$	33	
4	$4 \times 73 = 292$	2	44	
5	$5 \times 73 = 365$	$2\frac{1}{2}$	55	
6	$6 \times 73 = 438$	3	66	
7	$7 \times 73 = 511$	$3\frac{1}{2}$	77	
8	$8 \times 73 = 584$	4	88	
9	$9 \times 73 = 657$	$4\frac{1}{2}$	99	
10	$10 \times 73 = 730$	5	110	
11	$11 \times 73 = 803$	$5\frac{1}{2}$	121	

fore this half when added to the half in 3½, will make one complete Nakshatra; and the total complete Nakshatras arising out of the भशेष in this case are, therefore, not 3 but 4. Similarly in the case of the 9th and the 11th paksha the number of additional Nakshatras arising out of the भशेष is 5 and 6 respectively and not 4 and 5. So far as the *ûna* pakshas alone are concerned these exceptional Nakshatras arise out of the भशेष in the case of the 7th, the 9th and the 11th *ûna* paksha only. But as observed in the *remark* column, we have to add to the amshas in column 3 (a) the amshas at the end of the previous dozen pakshas; and then the said exceptional cases may occur oftener. For example take the 17th paksha. Here, we have one dozen *plus* five ऊन pakshas, and though the amshas for five *ûna* pakshas are 55 only, yet when 8 amshas of the previous dozen are added to it we have 83 amshas, or half a Nakshatra, and one amsha. This half of a Nakshatra when added to 2½ Nakshatras in column 3 (a), makes three complete Nakshatras in all. Therefore although the complete Nakshatras arising out of the भशेष for five pakshas is 2 (one half in 2½ being omitted), yet for 17 pakshas the number of completed Nakshatras is 3 and not 2. Briefly stated, we may say that (1) the number of complete Nakshatras, arising out of the भशेष is generally equal to the integral part of half the number of *ûna*-pakshas. But (2) an increase in the number of complete Nakshatra may take place when the number of amshas for the *ûna*-paksha (at the rate of 11 per paksha) either by themselves, or together with the amshas of the previous dozens (at the rate of 8 per dozen) exceed 62. This increase is generally never greater than one Nakshatra but in rare cases, e. g. in the 119th paksha, it may be equal to 2 Nakshatras. We can now see why the Vedânga has given a separate rule for calculating the moon's भशेष. The figures given in column 3 (a) and 3 (b) are practically the same as would be obtained by the application of the general rule for counting the भांशा's of the Moon at the end of a given paksha, laid down in the verse भांशाः स्युः etc. (R. 10; Y. 15) discussed in the beginning of this note. In column 3 (b) we have successive multiples of 11, or in other words एकादशगुणश्चोनः; and half a Nakshatra, which requires to be added in the case of a bright paksha (शुक्लेऽर्धं etc.,) will be found to be so added in the case of all uneven pakshas in column 3 (a). For the *ûna*-pakshas commence after a dozen and so all the uneven *ûna*-pakshas are bright. But there is an important difference between these two results. The verse भांशाः स्युः etc., gives us only such amshas as are in excess of the completed Nakshatras; while by calculating according to the rule ऊनस्थाने-द्विसप्तती: etc., we get the whole of भशेष, that is, the complete Nakshatras arising out of it, as well as the amshas in excess thereof. And as both these are required to be known for the purpose in hand, it was necessary to give a separate rule for calculating the entire भशेष of the Moon.

Let us now see how the Vedânga derives the number of the required *Kalâs* from this भशेष. *Immediately after the above verse*, the Rik recension has the following:—

त्र्यंशी भशेषो द्विवसांशभागश्चतुर्दशस्याप्यपनीतभिन्नम् ।
भार्धेऽधिके चाधिगते परेंऽशे द्वावुत्तमैकं नवकैरवेयम् ॥ R.12; Y.27.

this order of the verses is not, however, preserved in the Yajus text; and both *B* and *S* have, in consequence, been led to interpret this verse independently. In their opinion it gives us a rule for finding the number of days, (*B*. Lunar; *S*. civil) required by the Sun to traverse one complete Nakshatra; and that the word चतुर्दश means the 14th day thereof. The word त्र्यंशी in the first half of the verse literally means "having a third (त्र्यंश)," "possessed of a third in addition" or, algebraically expressed, $a+1/3\ a$, and not three-fourths as *B* interprets it. *S* saw the difficulty and got over it by altering it into त्र्यश. Both have again to interpret भशेष to mean the remainder of a *day*, and not of a *Nakshatra* (भ) as it naturally means. As for the second line *S* changes it radically and *B* thinks that द्वौ in द्वावुत्तमैकं is the locative of द्वि which is believed by him, means a nycthemeron in the Vedânga. But this meaning of द्वि is so unnatural that without a special definition to that effect, it cannot, in my opinion, be accepted, even as a convention. Besides it is not necessary to strain the meaning in this way; and where it seems to be necessary I would adopt the reading द्यु instead, as Diksit and others have done in Y. 31, 37 and 38. Finally the rule obtained after using all these artifices and emendations is both incomplete and superfluous. Incomplete because it admittedly involves an error of about 16 *kalás* if *B*'s, and $13\frac{7}{9}$ daily amshas if *S*'s, interpretation be accepted. Superfluous because in another place (R. 18; Y 39) the Vedânga tells us that the Sun takes exactly $13\frac{5}{9}$ days to traverse a Nakshatra, and when we thus have the exact rule, there is no necessity of giving the same rule in an incomplete form. Both *B* and *S* have tried to answer this question in their own way. But the reasons adduced are so lame, that one is led to think, therefrom, that both *B* and *S* have put forward these interpretations, because nothing better suggested itself to them. *B* has even a lurking suspicion that his interpretation may be open to improvement. But if we follow the cue suggested by the order of these verses in the Rik. recension, and interpret the present verse as containing the second part of the rule for finding the Moon's *kalás*, in continuation of the first part contained in the r.e ऊनस्थाने द्विसप्तती: etc., above discussed, we meet with no such difficulties. Taking त्र्यंशी to mean "one and one third" and making it qualify both भशेष and दिवसांशभाग, I would, therefore, translate the first half of the verse thus :—" Four-thirds (*lit.* one and one-third) of भशेष, and of the दिवसांशभाग (will be equal to the *kalás* at the end) of the 14th (paksha), the fractions (भिन्न) being kept aside (that is, neglected)." What is meant by भशेष is already explained. The meaning of दिवसांशभाग, as well as the reason of the whole rule will be seen from the following explanation.

We have shown above that at the end of a given number of *ûna*-pakshas, the Moon will have traversed (1) complete Nakshatras equal to 14 times the number of pakshas *plus* (2) Nakshatras equal to the integral part of half the number of *ûna*-pakshas, and (3) under certain *condition* there is a chance of having one or rarely two more additional Nakshatras. The first half of the present verse deals with the first two cases and the second half with the third of them. The Moon, according to the Vedânga (R. 18 ; Y. 39), takes one day *plus* 7 *kalâs*, to do a Nakshatra ; and at this rate, 14 days *plus* $14 \times 7 = 98$ *kalâs* are required to traverse 14 complete Nakshatras. But since we have here to find the expired time on the last day only, we omit the completed days from the above result and take into account only the fractional 98 *kalâs*, or the fractional parts of a day in that result. Thus, so far as the first of three above-noted cases is concerned, we may say that at the end of each of the given *ûna*-paksha, the number of Moon's *kalâs* will be $= 14 \times 7$ or ·98 only. But the भशेष at the end of a paksha is $= 73$; and $73 + \frac{1}{3} \times 73$ is $= 97\frac{1}{3}$ or only a fraction less than 98. Therefore instead of beginning a second operation for finding the *kalâs* in question, we can derive the same from the भशेष, previously ascertained, by adding its third to it. This involves an error of $\frac{2}{3}$ of a *kalâ* per paksha ; but this small fraction may be neglected for the sake of ease and convenience. The भशेष with its third (त्र्यंशी), or, in other words, the four-thirds of the भशेष thus represents the number of *kalâs* at the end of the given *âna*-pakshas, so far as the completed Nakshatras at the rate of 14 per paksha are concerned. But out of the भशेष there also arise other complete Nakshtras equal to the integral part of half the number of pakshas; and the *kalâs* for these must also be computed. This is done by taking four-thirds of the दिवसांशभाग for each of the said Nakshatras. दिवसांशभाग with reference to the Moon means the seven *kalâs* or the fractional daily parts in excess of a *day* (दिवस) required by her to move through one complete Nakshatra. Thus, if there be one additional Nakshatra arising out of the भशेष, the fractional part in excess of a day would be 7 *kalâs* only, and so on in proportion; and, properly speaking, this fractional part of a day must alone be added to the result previously obtained. But the Vedânga asks us to take four-thirds of 7 or, omitting fractions, 9 *kalâs* instead, with a view to compensate for the error introduced in our operation by taking $97\frac{1}{3}$ for 98 in the first case. Practically we have thus to take nine and not seven *kalâs*. Some may prefer to interpret दिवसांशभाग in itself as meaning nine, as the Sun traverses 9 amshas per tithi. However, since we are speaking here of the Moon, I prefer to take, त्र्यंशी with दिवसांशभाग and thus make it equal to nine. Combining these two parts of the rule, we can easily calculate the time of the Moon's entry into the last parvan Nakshatra at the end of any paksha, when no further additional Nakshatras arise out of the भशेष. Thus if the given number of pakshas be 14, we split it into one dozen and two *úna*. For the dozen we have 19 *kalás*. For the two *únas* the भशेष is 2 × 73 =146, four-thirds of which is = 194 (the fractional part being omitted). In addition to this we must take four-thirds of दिवसांशभाग; and since half of two (*únas*) is one, there is only one additional Nakshatra, and in consequence only one दिवसांशभाग, the four-thirds of which is 9 (omitting fractions). Therefore the total number of *kalás* at the end of the 14 pakshas is = 19 + 194 + 9= 222. Similarly for the 16th paksha we take 19 *kalás* for the dozen and proceed for the 4 *úna*-pakshas as follows. The भशेष for 4 *úna*-pakshas is =4 × 73 = 292; and its four-thirds = 389. In addition to this we must take four-thirds of the दिवसांशभाग; or in the present case 9 × 2. Therefore the total number of *kalás* at the end of the 16th paksha is=19 + 389 + 18 = 426. And so on for other similar pakshas.

But the matter does not end here. The case of further extra Nakshatras arising under certain circumstances, that is, the 3rd of the above noted cases, is yet to be provided for; and this is done by the second half of the verse. The traditional reading of this half verse is:—

भार्धेऽधिके चाधिगते परेंऽशो द्वावुत्तमैकं नवकैरवेयं ॥

The last line is obviously a little corrupt. I, therefore,

read द्वावुक्तमेकं for द्वावुत्तमैकं, and नवकैरवेत्यं for नवकैरवेद्यं; or in the latter case we may also read नवकैश्चवेद्यं, the meaning being the same in either case as इ with अव् has the same sense as विद् to know. Thus read the verse means—"And when the *next* amsha amounts to half a Nakshatra or more the two (*united*) is said to be one, and should be counted (*lit.* known) by a group of nine (*kalás*)." A reference to the Table II, given before, will clearly show, what is meant by the rule. In the column 3 (a) of that table we have the number of Nakshatras arising out of the भशेष and in the next part of the same column, that is, in 3 (b) we have the amshas in excess thereof. The word पर in the verse refers to these amshas; which, as previously stated, may also be calculated by the general rule कादशगुणश्चोन: शुक्लेऽर्धं etc., (R. 10; Y. 15). So long as these amshas are less than 62 there is no chance of their affecting the number of Nakshatras in column 3 (a) of the table, whether that number be an integer, or an integer plus one half. But such is not the case when the amshas are equal to or exceed 62. We are, therefore, told that when these amshas are equal to or exceed 62, they must be united with the previous Nakshatra number after thus blending the two into one and the *kalás* should be calculated at 9 per each of the complete Nakshatras so obtained. For instance let the number of pakshas be 19, or one dozen plus 7. Here we take 19 *kalás* for the dozen; and proceed to calculate the *kalás* for the seven *únas* as follows:—The भशेष for seven *únas* is $= 7 \times 73 = 511$, the four thirds of which is $=511 + 170=681$ (omitting fractions). The number of additional Nakshatras is $\frac{1}{2} \times 7 = 3\frac{1}{2}$; and according to the general rule one would get only 3 additional complete Nakshatras. But the amshas corresponding to the 7th *úna* are 77, or half a Nakshatra *plus* 15. Therefore, combining this half a Nakshatra with the previous $3\frac{1}{2}$ we get 4 complete Nakshatras. This at the rate of 9 *kalás* gives us $9 \times 4 = 36$ *kalás*. The total number of *kalás* at the end of 19th paksha is thus $= 19 + 681 + 36 = 736$; or deducting 603 *kalás* of a complete day $= 133$. To state the result fully the time of the Moon's entry into the last parvan Nakshatra at the end of 19 pakshas is 133 *kalás* of the parvan day. Another example of the application of the rule is where the amshas of the *úna*-paksha, are by themselves less than half a Nakshatra but exceed it when combined with the amshas at the end of the previous dozen pakshas. For example, suppose the number of pakshas be 17 or one dozen plus five *úna*. Here for the dozen we take 19 *kalás*. For the five *úna* the भशेष is $5 \times 73 = 365$, the four-thirds of which is $= 365 + 121 = 486$ (omitting fractions). The Nakshatras arising out of the भशेष are one half of five *únas*, or $2\frac{1}{2}$ only; and the amshas corresponding to five ऊन are 55 only, [See Table II column 3 (*b*)], One may thus suppose that there are only two completed Nakshatras in this case. But the amshas, at the end of previous one dozen pakshas, are = 8 ; and 8 added to 55 makes 63, or one amsha more than half a Nakshatra. Therefore adding this one half to 2½ we get 3 complete Nakshatras, which, at the rate of 9 *kalás* per each, give us 27 *kalás* ; and the total number of *kalás* for the 17th paksha are 19 + 486 + 27=532. Or suppose that we require the time of the Moon's entry into the last parvan Nakshatra at the end of 92 pakshas. Here 92 is equal to 7 dozens + 8 *únas*. The *kalás* for 7 dozens are = 7 × 19 = 133. The भशेष for 8 *únas* is 73 × 8 = 584, the four-thirds of which is 584 + 194 = 778. The extra completed Nakshatras will ordinarily be one half *úna* pakshas or 4. But the amshas for the 8th *úna* are = 88, with 8 × 7 = 56 for the previous dozens ; and the two added together become equal to 144 amshas or 1 Nakshatra and 20 amshas. Therefore we must here take 5 completed Nakshatras and their *kalás* would be 5×9=45. Thus the total number of *kalás* is = 133+778+45=956, or deducting 603, equal to 353 only. Therefore the Moon in this case will enter the last parvan Nakshatra, after 353 *kalás* of the parvan day have elapsed.

I have discussed at some length the meaning of the last two verses, because their right significance, according to my view, has not, as yet, been pointed out by any one. But though the explanation be lengthy, yet the rule itself, as will be seen from the examples above worked out, is quite simple. We first count the *kalás* of the dozens at the rate of 19 per dozen and then proceed to deal with excess (*ûna*) pakshas. The rule for the latter purpose may be generalized as follows. If x be the number of *ûna* pakshas let a be equal to integral part of $x/2$; Then $73x$ will be the भशेष and $4/3\ (73x) + 9a$ will represent the *kalâs* at the end of the given *ûna*-paksha, provided that when the amshas corresponding to the *ûna* paksha either by themselves, or combined with the amshas for the previous dozen, are equal to or exceed half a Nakshatra, the value of a will receive a corresponding increase. We have noticed above the Vedânga artifice of taking four-thirds of 73 instead of 98, and compensating for the error by taking four-thirds of 7, that is, 9 instead of 7, to calculate the *Kalâs* from the भशेष previously determined. The following table will show to what extent the result so obtained deviates from the one calculated strictly according to the accurate Vedânga elements. In column 1 is given the value of x or the number of the *ûna* pakshas; in column 2 the value of a, as well as the increase it receives under the circumstance noted above. Column 3(a) gives the value of भशेष or $73x$, and 3(b) of $9a$; and their total is given in column 3(c). Column 4 gives the number of *kalâs* if we take 98 instead $4/3 \times 73$ and 7 instead 9, or calculate according to the formula $98x + 7a$; and the last column shows the error introduced, the sign + or − respectively showing that the calculated result is greater or less than the actual by the number following that sign.

TABLE III

The value of x or the No. of ज्ञ pakshas.	The value of a with its increase.	The number of *kalás* according to the Vedânga rule :— $\frac{4}{3} \times 73x + 9a$ = Total *kalás*.			The value of $98x + 7a$.	Difference between 3c & 4; or error + or —.
1	2	3			4	5
		a	b	c		
1	0	73 + 24 = 97	0	97	98	—1
2	1	146 + 48 = 194	9	203	203	0
3	1	219 + 73 = 292	9	301	301	0
4	2	292 + 97 = 389	18	407	406	+1
5	2	365 + 121 = 486	18	504	504	0
6	3	438 + 146 = 584	27	611	609	+2
7	3 + 1	511 + 170 = 681	36	717	714	+3
8	4	584 + 194 = 778	36	814	812	+2
9	4 + 1	657 + 219 = 876	45	921	917	+4
10	5	730 + 243 = 973	45	1018	1015	+3
11	5 + 1	803 + 267 = 1070	54	1124	1120	+4

The table is calculated for the *úna*-pakshas only; and to get the actual results, the amshas as well as the *kalás* at the end of the previous dozen pakshas will have to be added to the figures in the Table. But the latter does not affect the error, except when the value *a* is increased thereby, in which case the error would increase by 2. For example, this will happen when the number of the total pakshas is 17; because in that case 8 amshas of the previous dozen when added to 55 of the five *úna* give rise to an additional Nakshatra, thus increasing thereby the value of *a* by 1. If we except these cases, the error, as will be seen from the above Table, never exceeds 4 *kalás*, being actually zero in three cases. But taking the worst or the exceptional case noted above, the error may, at best, in a few cases, be equal to 6 *kalás*. My interpretation of the above verses, cannot, therefore, be objected to, on the ground of this error, especially, as the only alternative interpretations hitherto suggested, involve a much larger error, and give a superfluous rule in addition. It should further be noted that the error in the number of *kalás* grows to 4 in the 9th and 11th *úna* pakshas for the compensating term becomes a little too much in these cases. But soon after, that is, at the end of the 12th paksha, we make a new start with the *kalás* of the next dozen pakshas, which are not affected by this error. In other words, the error is thus corrected at the end of every dozen pakshas or soon as it grows to 4 *kalás*.

The next verse in the Rik. recension closes this subject in the following manner :—

पक्षात्पंचदशाच्चोर्ध्वं तद्भुक्तमिति निर्दिशेत् ।
नवकैरुद्धृतोंऽशः स्याद्नांशं द्व्यधिकेन तु ॥ R. 13.

The first half of the verse does not occur in the Yajus recension and the second is only partially reproduced in Y. 16. The preceding verse त्र्यंशी भशेषो etc., gave us a rule for finding the time of the Moon's entry into the last Nakshatra at the end of the 14th pakshas and recorded the changes to be made therein in certain exceptional cases. Now the present verse informs us that "from the 15th paksha and onwards one should indicate that (which is so calculated) as the (*kalás*) elapsed (at the time of the Moon's entry into the last parvan Nakshatra)". In other words the complete rule is to be applied to all the pakshas after the 14th. The rule and its application being thus fully stated, the second half of the verse briefly states the rates of increase involved in these operations. ऊनांशं is evidently a mistake for ऊनांशो ; and making this correction I thus translate the second line :—" Nine is the increment in amshas, and additional two, that is, 11 is the increment for the *ûna* (pakshas)." Here we may take 9 as referring to the increase in solar amshas per day, or better still to " the group of nine " mentioned in the previous verse (R. 12). But the *ûna* eleven obviously refers to एकादशगुणश्चोन: in Rik 11 (Y. 19); and it is not fair in my opinion to twist the meaning of द्व्यधिकेन

and make it seven instead of eleven. It is true that in Y. 16 we have ऊनः सप्तगुणो भवेत्—the *ûna* is sevenfold. But there it refers to a different matter altogether, viz., to the calculation of the Moon's position in time and gives us the increment for the odd (ऊन) tithis, in excess of 15 tithis or a paksha (R. 21; Y. 21). The Rik and the Yajus verses undoubtedly refer to the same *general* subject, viz., of increments. But they need not be identical on that account. The Yajus omits to mention the increment in the case of the *ûna* pakshas, while the Rik does not give us the *ûna* amshas in the Moon's case. That is the difference between the two; and whatever its cause may be, we should leave it as it is, without labouring to seek a correspondence of the two by distorting the ordinary meaning of words. For, not only in this, but in other respects also, the two Vedânga texts, if taken singly, are found to be incomplete. I am also of opinion that in Y. 16, the reading आवापस्त्वयुजेऽर्धस्यात् is preferable to आवापस्त्वमुजे द्वौस्यात् (*b*) or युस्यात्, as *S* would have it. It refers to the increment of half a Nakshatra in the case of a bright or the uneven paksha after the heliacal setting of the Moon. (R. 11; Y. 15). *B*'s interpretation of Y. 16 and the Rik verse now under consideration (पक्षात्पंच° etc.,) is, no doubt, highly ingenious. But unfortunately it is not consistent with the natural meaning and construction of the verses; while the idea of rejecting quarters of the whole circle of 124 daily amshas, seems to me to be more imaginary than real. Nor can I accept *S*'s emendation of द्व्यधिकेन into द्व्यद्विकेन. Both have had again to give up the natural meaning of the words पक्षात्पंचदशात्. But there is nothing strange in these make-shifts; for such were needed to make the verses intelligible so long as it was not perceived, that Rik 10-13 (both inclusive) formed a connected set of verses, and that though the Yajus text did not show it, the original order of the verses was happily preserved for us in the Rik text.

To ascertain the time of the Moon's entry into the last parvan Nakshatra at the end of any given number of pakshas when the said number is not exactly divisible by twelve, is a rather complex operation and, therefore, the subject could not be finished in a single verse. But not so with other cases. After determining the time (in *Kalâs*) of the Moon's entry into the last parvan Nakshatra, the next step is to ascertain the time of her entry into the successive Nakshatras on the successive tithi days. This is a simple operation. The Moon takes one day and 7 *kalâs* to do a Nakshatra. Therefore if the time of her entry into the last parvan Nakshatra is known, we have simply to add 7 *kalâs* to it to find the time of her entry into the next Nakshatra, on the next tithi day, and so on. This rule is given in the following verse:—

या: पर्वभादानकलास्तासु सप्तगुणां तिथिम् ।
प्रक्षिपेत् कलासमूहस्तु विद्यादादानकीः कलाः ॥ **R. 21; Y. 21.**

The verse is correctly interpreted by *B* and also by

Mr. Diksit before him. *S* tries to improve on this, but his refinements are, in my opinion, uncalled for. I quote the verse here, not for the purpose of discussing its meaning; but more to point out that the words पर्वभादानकला: at the beginning of this verse distinctly presuppose a rule for finding the *kalás* of the Moon's entry into the last parvan Nakshatra, and that no such rule is to be found in the Vedânga, unless the verse त्र्यंशा भशेषो etc. (R. 12; Y. 17) be interpreted in the way proposed by me above.

Hitherto we have discussed the verses giving rules for ascertaining three lunar positions, viz, (1) her parvan position in space, (2) the time of her entry into the last parvan Nakshatra, and (3) that of her entry into the tithi Nakshatra. Her tithi position in space, or in other words, her tithi Nakshatra has now to be ascertained. The verse which gives a rule for the purpose is contained not in the Rik but only in the Yajus text. It runs as follows:—

तिथिमेकादशाभ्यस्तां पर्वभांशसमन्विताम् ।
विभज्य भसमूहेन तिथिनक्षत्रमादिशेत् ॥ Y. 20.

Grammatically there is no flaw in the verse, and read straight off, it means.—"One should indicate the tithi Nakshatra by multiplying the tithi by 11, adding (to it) the Nakshatra amshas of the parvan, and dividing (the sum) by the total number of the Nakshatras (that is, 27)." *B* seems to have translated it correctly. But his explanation shows that he has misunderstood the rule as well as its reason. *S* was, therefore, justified in looking for another explanation; but, as usual, he tries to improve only by ingeniously changing एकादशाभ्यस्तां into अर्कदशाभ्यस्तां and even then he is unable to show that the verse gives us the exact number of amshas of the tithi Nakshatra in question, for his result falls short by 9/15 of an amsha per tithi. He has also to interpret भसमूहेन as meaning 124 instead of 27 as it naturally means. If we, however, disabuse our mind, of the idea that the rule gives us the amshas of the tithi Nakshatra and not merely the Nakshatra itself, all these difficulties at once disappear. The Vedânga divides a Nakshatra into 610 *kalâs* of which the Moon traverses 603 in a civil day, or $593\frac{17}{62}$ in a tithi. Leaving the deficiency to be separately collected, the Moon, we might safely say, changes her Nakshatra on every tithi day; for, as the number of tithis does not exceed, error would never amount to a full Nakshatra in such a calculation rough as it is. If the Moon's parvan Nakshatra be known, we have, therefore, simply to take the next Nakshatra for the next tithi and so on. The rule is simple enough. But simple or otherwise, it has to be derived from the parvan amshas, according to the method, generally followed in the Vedânga; and this is what the present verse directs us to do. The Jâvâdi list of Nakshatras in the Vedânga is so framed as to secure a constant equivalence between the amshas and the Nakshatras at the end of each parvan (R. 15; Y. 17). If the parvan amshas are less than 27 they directly indicate the parvan Nakshatra in the Jâvâdi order, while if the amshas are greater than 27, the remainder left after dividing them by 27, will indicate the parvan Nakshatra according to the Jâvâdi list. Now as a necessary result of the principle adopted in framing the Jâvâdi list, the Nakshatras taken in their natural order are separated by 11 places in the Jâvâdi list. Thus Aswini is in the first place in the Jâvâdi list, Bharani in the 12th, the Krittikâs in the 23rd, Rohini in the 34th, or *deducting* 27, in the 7th place; and so on. This explains the reason of the present rule. For the parvan amshas, calculated according to R. 10 (Y. 15) and divided by 27, when they exceed that number, indicate the Moon's parvan Nakshatra according to the Jâvâdi list. But on the next tithi day the Moon is in the next Nakshatra and so on in succession; and successive Nakshatras are represented in the Jâvâdi list by successive multiples of 11, divided by 27 when the multiple is greater than it. Therefore, if we add 11 per tithi in succession to the parvan amshas, the sum of *both* divided by 27, when it exceeds that number, will represent the successive tithi Nakshatra of the Moon. It is needless to say that the Vedânga generally indicates the Nakshatras by a reference to the Jâvâdi list, and when such is not the case some express word is used to mark the exception, as in the verse next here-in-after discussed.

So far we have examined the verses containing rules for determining the Moon's various positions. Let us now see how the Vedânga determines similar positions of the Sun. Rik. 10 (Y. 15), explained in the beginning of this note, enables us to determine the parvan position or the *bhâmshas* of the Sun at the end of a given parva and the following Yajus verse tells us how to find, therefrom, the *tithi* position of the Sun :—

एकादशभिरभ्यस्य पर्वाणि नवभिस्तिथिम् ।
युगलब्धं सपर्व स्याद्वर्तमानार्कभं क्रमात् ॥ Y. 25.

" Multiply the (number of the elapsed) parvans by 11 and the (current) tithi by 9. What is obtained by their sum, together with the (number of) parvans, would be current solar Nakshatra in (natural) order (from Shravishthâ)." The late Shankar Bâlkrishna Diksit was the first to explain this verse. But *B*, not being previously acquainted with Mr. Diksit's book, tried to give a different explanation. Upon this, *S*, following Diksit, pointed out the defective character of *B*'s explanation and the latter has now acknowledged it. There remains, therefore, little to be said in this connection ; except that *B*, in my opinion, justly objects to *S*'s (or rather Diksit's) explanation of युगलब्धं in the second half of the verse. युग, used as a numeral, may mean 2 or 4, but not 124 ; nor is it necessary to twist the meaning in this way, when the ordinary sense of " addition " suits the context very well. No one need tell us that in dealing with the Nakshatra-amshas, one must reject complete Nakshatras, or in other words, any multiples of 124, when the total amshas exceed that number. But if a Vedânga authority is needed for the purpose it will be found in Y. 12, as will be shown later on. Another fact to be noted, as expressly observed by *S*, is the propriety of the words सपर्वं and क्रमात्. It is not the amshas (divided by 27 if necessary), that here indicate the Nakshatra according to the Jâvâdi list, as in the case of the previous verse. But it is the number of parvans increased by the number of complete Nakshatras arising out of the amshas, that now indicates the Nakshatra, and that too, not in the Jâvâdi but in the natural order beginning with Shraviṣthâ. As the Sun changes his Nakshatra once in $13\frac{5}{9}$ days, we have in this case to calculate the change in the Nakshatra amshas only. The rule for determining the daily time-amshas at the end of a tithi, or in other words, for finding the time of the day when a tithi ends, is given in R. 20 (Y. 22). But, as there is no dispute about its meaning, it is not necessary to discuss the verse in this place.

The next question is to find out the time elapsed between the Sun's entry into the last Nakshatra and the end of a parvan or tithi. According to the usual method of the Vedânga we expect to find two verses dealing with the subject—one for the parvan and the other for the tithi calculation. But in the present case one rule is sufficient for both these purposes. The Sun takes $13\frac{5}{9}$ days to traverse a Nakshatra of 124 amshas; or, in other words, the Sun moves through 9 Nakshatra-amshas in a single *tithi* (R. 24, Y. 42). Therefore the

Nakshatra-amshas of the sun divided by 9 will at once give the tithi-periods elapsed since his first entry into the said Nakshatra. I use the words tithi-periods advisedly. For a tithi does not generally begin with the Sun's entry into a Nakshatra; and all that we are entitled to say is that the quotient obtained by dividing the solar amshas by 9, represents periods of time, each of which is equal in length to a *tithi*. But it is not convenient to measure time, in this case, by such tithi-periods. We have to convert them into days. This can be done as follows:—We know from the Vedânga that whereas a day contains 124 amshas, a tithi contains only 122; or that one tithi is equal to one day *minus* 2 daily amshas. Therefore, the above mentioned tithi periods will be equal to as many days *minus* daily amshas equal to double the number of the said tithi-periods. Expressed algebraically, if n be the number of amshas traversed by the Sun at a particular time, then $n/9$ is the number of the tithi periods elapsed between the Sun's entry into the Nakshatra and the time in question; and that $n/9$ tithis are $= n/9$ days $- 2n/9$ (daily) amshas. For example, the Sun's Nakshatra amshas at the end of the 11th parvan are 121; therefore the time elapsed between his entry into the last parvan Nakshatra and the end of the said parvan is $= 121/9 = 13\frac{4}{9}$ tithi-periods or $13\frac{4}{9}$ civil days *minus* $26\frac{8}{9}$ (daily) amshas. Or we may proceed to calculate in a silghtly different way. We know that the Sun traverses 124 amshas (one Nakshatra) in $13\frac{5}{9}$ days. Therefore, by a simple rule of three we can directly obtain the time, in days, required by the Sun to traverse 121 amshas. The result is the same in either case. *B* and *S* both agree that, in one form or another, this rule is contained in the following verse :—

सूर्यर्क्षभागान् नवभिर्विभज्य शेषद्विरभ्यस्य दिनोपभुक्तिः ।
तिथिर्युता भुक्तिदिनेषु कालो योगो दिनैकादशकेन तद्भम् ॥ Y. 26.

But they differ widely in their detailed explanations. *B* takes शेषं to mean the whole quotient including the remainder, though the word ordinarily denotes the remainder only. He has further complicated the problem by making it too general. He thinks that the Vedânga here gives us a rule for finding the elapsed time not at the end of a parvan or tithi but generally at any time during any tithi day. His interpretation of दिनैकादशेन is, however, at once simple and reasonable. *S* proceeds differently. He rightly understands शेषं to mean the remainder and not the whole result as *B* does. But, supplying a lot of his own words, he interprets the verse to mean :—" Divide the solar amshas by 9, double (the whole including the remainder), the difference (शेष) (after subtracting the latter, in amshas, from the former as days) is the time in days elapsed." Here we have not only to supply all the words bracketed, in a somewhat artificial manner, but, if the verse be so interpreted, the second half of the line becomes perfectly redundant, as the interpretation gives us, at once, the whole formula, viz., $n/9$ days minus $2n/9$ amshas. *S* tries to get over this difficulty by altering the reading of the second half and showing that this part of the verse refers us to some other rule for finding out the time previous to the Sun's entry into the Nakshatra under consideration. This, in my opinion, cannot be regarded as satisfactory. We naturally expect to find दिनैकादशकेन etc, made a part and parcel of the rule for ascertaining the time of the Sun's entry into the given Nakshatra; and this expectation is defeated if the first half of the verse is made to yield the whole formula as *S* has done. The course adopted by *B* is certainly more preferable to this. But, as shown above, *B*'s explanation is not also free from defects. Under these circumstances I would suggest the following modification thereof. I would first restrict the scope of the rule. The Vedânga calculates the Sun's amshas only in two particular cases :—(1) at the *end* of a parvan (R. 10, Y. 15) and (2) at the *end* of a tithi (Y. 25). There is, therefore, no reason to suppose that the word सूर्यर्क्षभागान् in the present verse refers not to either of these two cases, but to a still more general one. The next step is to ascertain the meaning of the word दिनोपभुक्ति in the second line, or its synonym भुक्तिदिन in the third. In my opinion दिनोपभुक्ति does not here mean "the total time elapsed" as *S* interprets it, nor do I understand how *B* translates it as "the total of the Sun's motion (in the equinoctial circle) per lunar day." (Perhaps, he means "the total motion calculated at so much per lunar day "). The word appears to me to be here used in a somewhat technical sense. The relation between a tithi and a civil day is usually expressed in the Vedânga not in the form of a fraction $(\frac{122}{124})$ as we should now do; but by mentioning along with the tithi the two (daily) amshas not covered by it. In other words, these two amshas belong not to the tithi, but to the day on that tithi; and may properly be termed the दिनोपभुक्ति (that is the day enjoyed, or day covered parts) *of*, or *with reference to the tithi* in question. Thus 15 tithis have a दिनोपभुक्ति of 30 daily amshas; and in stating the relation of 15 tithis to 15 days the Vedânga simply mentions this दिनोपभुक्ति side by side with the number of days, it being understood that it is to be subtracted from the days, to find the exact corresponding time in question. Such, at any rate, seems to me to be the method followed in the present verse; and taking तिथे: for तिथि: in the third line, I would interpret the verse as follows. The first line is quite simple. It asks us to "divide the solar Nakshatra amshas by 9." The integral quotient at once gives us the number of tithis or rather the tithi-periods. But as the Sun's Nakshatra-amshas are not always completely divisible by 9, there will generally be left some remainder; and as a fraction of a tithi is not generally spoken of as 'a tithi' it is necessary to say how the remainder is to be dealt with. The second line tells us what to do in such a case. "Multiply the remainder by two and it will be the दिनोपभुक्ति," not by itself alone but, says the third line, "together with the दिनोपभुक्ति of the tithi". The word 'tithi' does not here mean the current tithi of the day. It means the "tithi (periods) represented by the integral quotient"; and as the दिनोपभुक्ति of a tithi (in daily amshas) is equal to double the tithi (number), the Vedânga here practically asks us to double both the integral quotient and the remainder. But it is done in such a way as to elucidate the reason of the rule along with it. The word कालः in the third line may be construed with दिनोपभुक्तिः, or it may be taken with the fourth line to denote the total number of days elapsed. The meaning of the whole verse is not altered thereby. We have thus obtained the दिनोपभुक्ति of the integral and the fractional tithi periods. But without knowing the corresponding number of days the answer is incomplete. The last line of the verse, therefore, tells us how to find the total number of days corresponding to this दिनोपभुक्ति. It says that "the sum (that is, दिनोपभुक्ति in amshas *plus* the given number of solar amshas) is the time (in days) since (the sun's entry into) the Nakshatra, at the rate of eleven (amshas) per day". *B* has correctly explained the reason of this procedure. The दिनोपभुक्ति is equal to twice the number of tithis; and the solar Nakshatra amshas are equal to nine times the same number. The sum of the two (considered as *numbers* only) when divided by 11 (2+9), will, therefore, give us the *number* of tithis (fractions included), which is the same as the number of days; and, taken along with the दिनोपभुक्ति previously ascertained, we get a complete answer to our question. This is practically equivalent to saying that the whole quotient (including the fractional remainder) represents the total number of days corresponding to the दिनोपभुक्ति. But the Vedânga seems to have followed the indirect method in order to keep the uninterrupted continuity of the arithmetical operation. The final result obtained is thus the same as that obtained by *S* or *B*. In fact the interpretation here proposed is only a modification of *B*'s. But, in my opinion, it enables us better to keep by the natural construction and the meaning of the words in the text. Whether it actually does so or not, is for others to decide.

There is a verse in the Yajus text of the Vedânga (Y. 12), which has been interpreted nearly in the same way both by *B* and *S*. *S* does not include it among the nine verses mentioned by him in the preface to his Bhashya, as wrongly interpreted by *B*; while on the other hand *B* has observed that *S*, in putting forward a different interpretation of this verse, has simply sought "to draw a distinction without a difference". But as I have to propose a new interpretation altogether, it is necessary to examine, in this place, *B*'s as well as *S*'s interpretation thereof. The traditional text of the verse is as follows:—

दुहेयं पर्व चेत् पादे पादस्त्रिंशत्तु सैकिका ।
भागात्मनाऽवपृज्यांशान् निर्दिशेदधिको यदि ॥ Y. 12.

Both *B* and *S* read अधिकं यदि for अधिको यदि in the last line. But this is not absolutely necessary as even without this correction the words will have the same sense if we can construe यदि अधिकः (भागः) स्यात् (तर्दितं) निर्दिशेत्. The real difficulty does not lie in the last line, but in the first word of the first line. The rest of the verse is simple enough and read straight off may be thus translated :— "दुहेयं if a parvan is at a pâda. A pâda is thirty and one (amshas). One should indicate the excess, if any, after dividing the amshas by (all) the bhâgâs (amshas) themselves." Here we naturally expect दुहेयं to tell us what one should do "if a parvan is at a pâda." Or, in the terminology of grammar, दुहेयं appears to be the consequential clause depending on पर्व चेत् पादे. But there is no such verbal form as दुहेयं or even दुहेय in Sanskrit ; and, therefore, nothing can be 'milked' out of it so long as the word stands as it does. हेयं suggests the idea that something in neuter gender is here " to be abandoned or omitted." But what of दु prefixed to it ? Some suppose that the parvan itself is to be abandoned. But such is never the case in practice. In case of अधिमास we may say that the मास or two parvans are abandoned, that is, not included in the usual reckoning; but a single parvan is never so omitted. *B* skillfully tries to get out of the difficulty by changing दुहेयं into दुर्हेयं and supposing, that the amshas here spoken of are not the (parvan) Nakshatra-amshas, as the context plainly suggests, but the parts (amshas) of a day, he thus construes the verse :—पर्वं पादे दुर्हेयं चेत्(पादं) अवपृज्य अधिकं अंशान् निर्दिशेत्। पादः भागात्मना सैकिका त्रिंशत् स्यात्। He himself translates it as follows:—"If the (hour angle of the) parvan is not less than a quarter (of the equinoctial circle), the (latter) must be deducted from the former; and should the parvan exceed a quarter the remaining amshas be adopted. A quarter contains thirty one parts (bhâgâs or amshas)." Now in the first place दुर्हेयं is an unusual word; and secondly it means "difficult to be abandoned" and not "less than" as *B* understands it. अधिकं अंशान् is again ungrammatical. But these are not the only defects in *B*'s construction of the verse. His *anvaya* is extremely far-fetched and laboured. If the verse is read in its natural order we expect भागात्मना and अंशान् to go with अवपृज्य; and अधिकं with निर्दिशेत्. But *B* changes all this and tells us to take भागात्मना with त्रिंशत्, अवपृज्य with पादं understood (as if the author could not have said पादात्मना instead of भागात्मना), and अंशान् with निर्दिशेत्. The meaning of the verse, obtained after so much labour, is also not satisfactory. We are told that in the days of the Vedânga a nycthemeron (day and night) of 124 amshas was divided into 4 quarters of 31 amshas each, and what is still more important, the reckoning of the daily time stopped and recommenced at every quarter of 31 amshas, somewhat as we now do at 12 noon and 12 midnight. Therefore, the present verse of the Vedânga asks us to deduct a quarter or 31 amshas when the amshas of a day exceed it and name only the excess. Thus if an event happens after midday we are not to say that it occured at 40 or 50 amshas of the day but 40 - 31 = 9 or 50 - 31 = 19 amshas only. This meaning seems to be so out of place that one may very well ask if there be an example of it in the Vedânga. Yes, answers *B*; and points to *his* interpretation of Y. 16. But *B*'s interpretation of Y. 16 is unfortunately as far-fetched as that of the present verse. So, at best, we have a doubtful rule, supported by an equally doubtful example. One may, however, fairly say, that if the verse is not otherwise intelligible, there is no alternative, but to accept *B*'s meaning strained though it may be. This seems to be the view taken by *S*, who practically follows *B*, only proposing to read द्योर्ह्येयं instead of *B*'s दुर्हेयं, first because perceiving the right meaning of दुर्हेयं he saw that it was not the suitable word and secondly because he might have felt that some express authority was needed to hold that the amshas, mentioned in the verse, were the amshas of a day and not of a Nakshatra. *B* in reply calls this a specious emendation at once "artificial and unnecessary;" and so it might be, though I think otherwise, if *B*'s interpretation is on the whole to be accepted. But taking a hint therefrom, I proposed to read द्युहेयं for the meaningless and impossible दुहेयं and interpret the the whole verse in an entirely different way. The verse thus read will stand as follows :—

द्युहेयं पर्व चेत् पादे पादस्त्रिंशत्तु सैकिका ।
भागात्मनाऽवपृज्यांशान् निर्दिशेदधिकं यदि ॥

And taking the amshas to mean, as the context shows, the Nakshatra-amshas of a parvan, I thus explain the verse:—"A day, a nycthemeron ज्ञु should be abandoned or omitted." When? "If a parvan is at (that is, ends in) a pâda." What is a pâda? "A pâda is thirty and one (amshas)," says the second line. How are the (parvan) amshas to be counted? "One should indicate the excess, if any, after dividing the amshas by (all) the bhâgas (amshas) themselves (that is, by 124)." The *anvaya* of the whole verse thus becomes quite simple and natural; and an important rule is obtained therefrom. The second part of the rule requires little explanation. One need not go in search for an example to illustrate it. It is the actual procedure followed in calculating the parvan amshas according to the Vedânga. As directed in R. 10 (Y. 15) we multiply the number of parvans by 11, and then dividing the result by 124 (the total number of amshas in a Nakshatra) take the excess (without omitting quarters) to represent the Nakshatra-amshas at the end of a parvan. The rule had, however, to be given somewhere in the Vedânga, and the second half of the verse gives it to us in plain words. When the parvan amshas so calculated are equal to 31,—and this happens only once in a *Yuga*, viz, at the end of the 93rd parvan,—a civil day (a nycthemeron) is to be abandoned or omitted from our reckoning. In short, it is an extra or leap or intercalary day. This is the plain meaning of the verse; and the following explanation will disclose the importance and the necessity of this correction in the Vedâṇga calendar. The idea of omitting a day is not a new device of the Vedânga Jyotisha. As shown by me elsewhere, it is the basis of the *Utsarginam ayanam*, and is expressly mentioned in the Taittirîya Sanhitâ VII.5. 7. 1. In the case of yearly sacrificial *sattras* like the *Gavâm ayanam*, the *vishuvat* or the central day was also always omitted in counting the 360 days of the sacrificial year.

The astronomical elements on which the Vedânga rules are based, represent only the *mean* (मध्यम) motions of the Sun and the Moon. But the Sun's or the Moon's *actual* (स्फुट) position in the celestial sphere is not generally the mean, but a few degrees in advance or behind it. Besides this the Vedânga *mean* motions themselves are again not exact but only approximately correct. The Sun does not complete exactly five revolutions in a *Yuga* of 1830 days; nor is the number of lunar pakshas therein exactly equal to 124, as the Vedânga assumes it to be. At the rate of five revolutions per *Yuga* of five years, a solar siderial year becomes, according to the Vedânga, exactly equal to 366 days, whereas according to modern more accurate observations the same contains 365.25636 ...... (or roughly 365¼) days. This yearly error would amount to about one lunar month in 39.7, or, in round numbers, 40 years altering the Sun's position amongst the fixed stars by a month in advance. It is impossible that this could not have been noticed;

and the late Mr. Krishnashâstri Godbole thought the error was probably corrected by omitting one intercalary month in 40 years, that is, once in 8 *Yugas*; while Mr. Diksit believed that 35, instead of 38, intercalary months were inserted in 95 years, that is, in 19 *Yugas*. But whatever the method adopted might be, it was not necessary to speak of it in a book, professedly devoted only to the preparation of a five years (*Yuga*) calendar (R. 32;. Y. 5). In other words, it did not fall within the scope of the Vedânga Jyotisha and there is nothing surprising if Vedânga gives us no rule on the point. Not so with the Moon. The Vedânga lunar month (of two lunar pakshas) contains $\frac{1830}{62}$ or 29.5161290... days; whereas the average length thereof, according to modern research, is = 29.5305887.....days. The Vedânga month is thus *shorter* than the more accurate modern mean by ·0144597....of a day, which is equal to 20.82 minutes, = 8.719··· (Vedânga) *kalâs* or= 1.793...*amshas* of a day, (a day being made of 124 amshas as in the Vedânga). The error for a parvan or a paksha would be half of this; and at this rate it would amount to a day after 138 pakshas (69 lunar months), or about 53.8 (or in round numbers 54) ghatis (nâdikas) in a Vedânga Yuga of 124 pakshas. In a calendar prepared according to the Vedânga rules, the calculated full and new Moons would, therefore, fall behind the actual nearly by a day towards the end of a *Yuga*; and the Yajnikas, for whom the Vedânga rules were intended, could not have failed to mark it as they must have carefully watched the full and the new Moon, as actual celestial phenomena, owing to their sacrificial importance. Here was an error which the Vedânga calendar was bound to notice; for otherwise all the calculated full Moons for the rest of the *Yuga* would go wrong, thus rendering the calendar entirely useless. All the students of the Vedânga are, therefore, of opinion that this error must have been somehow or other provided for, though they have not been able to discover the specific way. As the error amounts to about 54 ghatis, that is, six ghatis less than a day, per *Yuga* Krishnashâstri Godbole thought that one more day was added to the second intercalary month of a *Yuga*, and that this correction was omitted at the end of every tenth *Yuga* to compensate for the excess of 6 ghatis included therein, (See page 32 of his pamphlet on *The Antiquity of the Vedas*, 1882). He went even so far as to predict that his suggestion about these corrections "would be found to be true as the careful study of the Vaidik and the post-Vaidik works would advance". The late Mr. Shankar Bâlkrishna Diksit, writing on the same subject, in his *History of Indian Astronomy* (p. 92), has further observed that though the Vedânga *Yuga* was made to consist only of 1830 days for facility of arithmetical calculation, yet *the full Moon must have been determined by actual observation*, or in other words, *the Yuga practically consisted of 1831, instead of 1830 days.*" (the italics are mine). The interpretation of the present verse, proposed by me above, is thus in full accord with the anticipations and observations of the previous students of the Vedânga.

The necessity of intercalating a day in a *Yuga* being thus established, the next question is as regards the exact time when the intercalation should be made. Krishnashâstri thought a day was probably added to the second intercalary month. But a little consideration would show that this could not have been the case. It was not a question merely of a day less or more in a *Yuga*, so that the extra day might be inserted at any time. The intercalation was needed to set aright the full Moon night; and the proper time for intercalating would be soon after or just when the calculated mean full Moon was observed to fall back a day earlier than the real *paurnimâ*, which must have been watched as an actual phenomenon. Here we must, therefore, compare the Vedânga mean full Moon with the one actually *observed* at the time. This cannot be accurately done, as there are no records of such observations; and it is doubtful if the best modern lunar tables would enable us to correctly ascertain the exact moment of the commencement of the actual phenomenon of the full Moon which occurred three thousand years ago, more especially as the exact date of the Vedânga Jyotisha still remains uncertain. But a good approximation can be made by comparing the Vedânga mean full Moon with the same, calculated according to the more accurate mean motion determined by modern research. According to the Vedânga a paksha contains

14 days (Nycthemera) and 94 amshas (one day being= 124 amshas); while according to modern research the average length of a paksha is 14.7652943··· days, which converted, for comparison's sake, into *amshas* is equal to 14 days and 94.896 *amshas* (one day = 124 amshas). Thus the Vedânga mean paksha is shorter than the real mean paksha by .896 of a day; and the end of each Vedânga parvan would occur earlier than the real mean by. 896 amshas of a day or by nearly 10 minutes or 4 *kalâs*, as stated previously. This difference accumulates as the number of pakshas increases and in course of time the calculated Vedânga full Moon must fall back in proportion. This is shown in the following table.* As the full Moon paksha is always represented by an uneven number in the Vedânga, these alone are noted in the table.

It is not necessary to consider the pakshas previous to 81, as the difference between the two results does not affect the number of the full Moon nights therein. The 81st is a bright paksha; and the pakshas, as well as the full Moon period thereof, end according to the Vedânga, when 1195 days and 50 amshas of the 1196th day from the beginning of the *Yuga* are elapsed. The 1195th night is, therefore, the mean full Moon night according to the Vedânga. But, according to the modern mean, the 81st paksha ends when 1195 days and 122 amshas (omitting fractions) of the 1196th day are elapsed, that

* See Table No. IV, p. 96.

TABLE IV

Serial No. of the pakshas.	Days and amshas at the end thereof according to the Vedânga.		The same according to the *modern* mean rate.	
	Days.	Amshas	Days.	Amshas.
81	1195	50	1195	122.616
83	1224	114	1225	64.409
85	1254	54	1255	6.202
87	1283	118	1284	71.995
89	1313	38	1314	13.788
91	1342	122	1343	79.581
93	1372	62	1373	21.374

is, only last 2 amshas of the 1195th night are covered by the mean full Moon. In other words, the calculated Vedânga mean full Moon now falls back not exactly one night but 2 amshas less; and as yet there was, therefore, no necessity of intercalating a day. The 83rd paksha, or the next mean full Moon night ends, according to Vedânga, after 1224 days and 114 amshas of the 1225th day are elapsed; that is, the Vedânga mean full Moon lasts till 52, (114–62), amshas of the 1225th night. The real mean full Moon, on the other hand, lasts till 64–2 amshas of the 1226th day, and therefore it begins at 2 amshas of the 1225th night. Practically the Vedânga mean and real mean full Moon may, therefore, be said to occur on the 1225th or the same night; and the same is the case with the 87th and 91st paksha. The error in the two mean full Moons causes, for the first time, a change of one full night at the end of the 85th paksha; but we may leave this case also, out of our consideration, since the actual place of the Moon in the heaven is not exactly the mean one, but differs from it by a few degrees on either side. There thus remain two cases, viz., of the 89th and the 93rd paksha, wherein, according to our mean calculations, the real full Moon falls on the night next to the Vedânga calculated one. Thus, in the case of 89th paksha, the Vedânga mean full Moon falls on the 1313th night, while the real mean full Moon begins 13 amshas later, so that it falls on the 1314th night; and same is the case with the 93rd paksha. The Vedânga mean full

Moon falls on the 1372nd night and ends by the evening of 1373rd day, so that not a trace of it falls on the 1373rd night according to the Vedânga, while the real mean full Moon occurs entirely on that night. Of these two cases the Vedânga has selected the latter for intercalation, either because it was a *Vishuvat* day, and it was an old practice to omit it in counting the days of the sacrificial yearly *Sattras* or, what seems more probable, because in these days it was at the end of the 93rd paksha that the error was, *by actual observation*, found too great to be any longer neglected, a fact which our calculations of the mean full moons is not likely to disclose to us; or it may be that no correction was made until one full moon was, as a matter of fact, actually observed to go wrong by a day. In any case the above calculation, of the mean full Moon though it is, sufficiently accounts for the selection of the end of the 93rd paksha for intercalating a day in a *Yuga*. According to the Vedânga calendar, it is the full Moon day of Kârtika in the fourth (*anuvatsara*), or what may now be called the leap year, of a *Yuga*. A glance at the table of parvan-amshas given in Mr. Diksit's book (pp. 77-78) will show—or it may be determined otherwise—that out of 124 pakshas of a *Yuga* only one—that is, the 93rd—ends in 31 amshas. There is, therefore, no ambiguity caused by the parvan being defined by mentioning its general amshas only, as the Vedânga here does. It has been already pointed out that it was necessary to omit this correction every fifth *Yuga*. But as this part of the rule did not fall within the scope of five years' calendar, like the solar corrections above mentioned, it might not have been included in the Vedânga.

But though the calendar may thus ostracise day, assigning no tithi or Nakshatra to it, the priest has to perform some sacrifice even on this outcast day; and one naturally asked what Nakshatra should be assigned to this day, for sacrificial purposes, especially as the tithi (lunar) Nakshatra changed from day to day. There are two courses open in this case, either (1) to treat this extra day as a part of the preceding parvan, or (2) to assign to it the Nakshatra in which the Moon may actually be observed to be on this day. These two courses are, in my opinion, described in Y. 14 by way of a corollary to the preceding verse as interpreted by me. The verse runs as follows:—

स्युः पादोर्ध्वं त्रिपद्यायास्त्रिद्व्येकेऽन्हः कृते स्थितिम् ।
साम्येनेंदोस्तृणोऽन्येतु पंचकाः पर्वसंमिताः ॥ Y. 14.

I adopt the manuscript reading पादोर्ध्वं for पादार्धं of Somakar and deleting the *anuswâr* in स्थितिं take स्थितिसाम्येन as one word or read स्थितेः instead. Thus corrected, the verse means:—" After the (last) pâda (of the previous parvan), the first, the second and the third (pâdas) of the Tripadyâ (the *pratipad*) would, for sacrificial purposes, be in the same position as the (actual) Nakshatra of the Moon (on that day). Others, however, (think) that the five (pâdas—the last of the parvan and all the four of the Tripadyâ) should be included in the parvan." *B* has for the first time correctly pointed out the meaning of Tripadyâ, as well as the practice of Yâgniks, "to regard the last *quarter* of the parvan and the first three *quarters* of the *pratipad* as forming the *Yagakâla* (sacrifice time)." These are exactly the four *pâdas* mentioned in the first half of the present verse. But *B* takes the verse as referring to any *pratipad* in general; and, adopting the reading पादार्धाः for पादार्धं, he divides the *pratipad* into 8 praharas or pâdârdhas and assigns 3 to the pratipad and includes the remaining 5 in the parvan. But I fail to see how the remaining five pâdârdhas of the *pratipad* can, in this case, be said to be included in the parvan, unless parvan be taken to mean the whole of the next fortnight. We have also to change तृणः into त्रयः make कृते govern स्थितिं in the accusative case instead of अन्हः in the genetive as it should naturally do, and finally suppose that a rule laid down in express terms only for the Moon's conjunction (इंदोः साम्येन–as *B* interprets it) is to be extended also to the full-Moon parvan. *S* follows *B* in general, but makes some further changes in the first half and interprets साम्येन to mean कारणेन which is quite unnatural. These difficulties are avoided if we take the verse as referring to the *pratipad*, at the time of intercalation, and not to any *pratipad* generally. But everything depends upon how we interpret Y. 12.

The last of the nine verses wherein *S* seriously differs from *B* is contained only in the Rik recension. It is as follows :—

श्रविष्ठाभ्यां गुणाभ्यस्तान् प्राग्विलग्नान् विनिर्दिशेत् ।
सूर्यान् मासान् षडभ्यस्तान् विद्याच्चांद्रमसानृतून् ॥ R. 19.

There is not much difference of opinion about the meaning of the second half of the verse. *B*, and Mr. Diksit before him, have interpreted it to mean—"Solar months multiplied by six should be known as so many lunar *ṛitus*." But the statement is, as pointed out by Mr. Diksit, only approximately true; for, according to the Vedânga, there are 67 lunar (siderial) months (Y. 30) and consequently 67×6 and not 60×6 lunar *ṛitus* in 60 solar months of a *Yuga*. *S*, therefore, changes सूर्यान् into स्वार्क्षान् and interprets the line thus:—"The lunar ṛitus are six times the (number of) lunar siderial months." This makes it a pure definition without instituting any comparison between the lunar ṛitus and solar months; and the objection, noted by Mr. Diksit, no longer exists. But this is not of much consequence. The real dispute is about the meaning of the first half of the verse and especially of the words गुणाभ्यस्तान् and प्राग्विलग्नान् therein. प्राग्विलग्नान् literally means "attached, or affixed to the east," and गुण, used as a numeral, ordinarily denotes 3, गुणाभ्यस्तान् thus meaning "multiplied by three." But *B* takes गुण to mean 8 in this place, and अभ्यस्त to signify "heaped over each other" or, in other words, "added to each other." With the help of these distortions he thus translates the first half verse,—"the 8th (group of stars) from Shravishṭhâ should be designated the equatorial (lit. the east-affixed) Nakshatra." The eighth Nakshatra from Shravishthâ is the Kṛittikâs, and *B*, in support of his interpretation, quotes the now well-known passage from the Shatapatha Brâhmaṇa (II, 1. 2. 1-3) which states that "The Kṛittikâs never deviate from the East." But I fail to see the relevency of the quotation. Whether the Kṛittikâs were or were not then regarded as the eastern stars is not the point at issue. We have, here, to see whether the supposed reference to the eastern position of the Kṛittikâs in the Vedânga follows from the natural interpretation of the present verse; and this question cannot, in my opinion, be answered in the affirmative. *S* was, therefore, right in seeking for another explanation; but the alternative proposed by him is not less objectionable. Taking the word लग्न, in प्राग्विलग्नान्, in its technical sense viz., and as denoting the part of the celestial zodiac in contact with eastern horizon at particular time, he interprets the Verse so as to give us a rule for finding the इष्टकाल, from which the लग्न may then be subsequently determined. But one may fairly ask that if the ascertainment of लग्न be the real subject of the verse, why is only a subsidiary and not the main rule given. Besides, *S* has to change गण into गण and then taking भ understood before गण eventually to interpret it to mean भगण = 27. श्रविष्ठाभ्याम् has again to be understood in the sense of "the rising of Shravishṭhâ." Of course, *S* has given us a mathematical demonstration of the rule which he thus derives from the verse. But a mathematical proof, howsoever rigorous it may be, is of little value if the meaning proposed does not naturally follow from the verse. Mr. Diksit has not translated the first half of the verse. But in several places of his book he has thrown out certain suggestions regarding its meaning, which deserve to be noticed. He has shown that before the introduction of *Râshis* and along with it the twelve *Lagnas*, the number of *Lagnas* was nine, each consisting of three Nakshatras, (pp. 97, 99 and 519). If so, one may interpret the verse as meaning that "one should indicate the *lagnas* by the (successive) multiples of three (counted) from Shravishṭhâ," without straining the meaning of any word therein. But even this meaning is merely conjectural; and in the absence of any further accurate information about the number and meaning of *lagnas* in the pre-Râshi period of Hindu astronomy, Mr. Diksit was right in leaving this part of the verse unexplained. The Vedânga rules were intended for ordinary priests; and it is not reasonable to assume that they were originally expressed in any but the simplest language and in the simplest manner, consistent with the nature of the subject. True that the Vedânga has long been a conundrum to us. But this was due partly to its present fragmentary character, partly to the corrupt state of the text and its technical nature and partly to our ignorance of the ancient astronomical methods. Thanks to the labours of Thibaut, Diksit, Bârhaspatya and others, these difficulties have been almost overcome. But, still, if a verse in the Vedânga fails to yield any intelligible meaning, except by violating the natural construction and meaning of the words used, we may be sure to have missed its true import; and the safest course to follow in such case is simply to record our suggestions, if any, and to leave the verse to be finally deciphered by future workers in the field, rather than try to give, by hook or crook, to our work a semblance of completeness it has not really attained. For, inspite of the great progress already made, the last word on the Vedânga is, in my opinion, yet to be uttered.

A MISSING VERSE IN THE SANKHYA KARIKAS.

(1915)

A
MISSING VERSE IN
THE SANKHYA-KARIKAS.*

The *Sâṅkhya Kârikâs* by Ishvarakṛishṇa is, in my opinion,the oldest work now available on the Sâṅkhya philosophy. Some regard *Sâṅkhya-pravachana-Sûtras* to be older. But if we compare some of the *Sûtras* with the corresponding *Kârikâs* they will be found to be almost the same word for word. Thus, for instance, *Sûtras* i. 140-144 exactly correspond to the 17th *Kârikâ*. Now the *Sûtrâs* are in prose and the *Kârikâs* in the *Âryâ* metre; and as the prose *Sûtras*, when read together, cannot be naturally supposed to make an *Âryâ*, it is but fair to infer that the *Sûtras* are obtained by splitting up a *Kârikâ* into so many prose sections. In other words the *Kârikâs* are older than the *Sûtras*.

As an old work on Hindu evolutionary philosophy, the *Sâṅkhya-Kârikâs* have received considerable attention at the hands of Western scholars and have been translated into Latin, German, French and English. The first English translation was by Colebrooke ; and this together with the translation of Gauḍapâda's *Bhâshya* and notes etc. was published at Oxford by H. H. Wilson in 1837. This edition is now out of print. But a reprint of the same has been published by the late Mr. Tukaram Tâtyâ of the Theosophical Society in Bombay in 1887.

*Contributed to the SANSKRIT RESEARCH Vol. I, No. 2, published in October 1915.

Recent Indian editions of the *Sâṅkhya-Kârikâs* are more or less a reproduction of the Oxford edition.

It is questionable whether the *Bhâshyakâra* Gauḍapâda is the same person as the grand preceptor of the great Shaṅkarâchârya. For, it does not seem probable that an *Advaita* Vedântist would care to write a *Bhâshya* on a *Dvaita* system of philosophy. We learn from Buddhistic works that Ishvarakṛishṇa was the literary and philosophical opponent of the preceptor of the great Buddhist scholar Vasubandhu of about the fourth century after Christ; and the *Kârikâs* with a commentary thereon were later translated into Chinese by Paramârtha. This Chinese translation of the *Kârikâs* with a commentary on it has been now rendered into French by Dr. Takakusu and published in the *Bulletin de l'E'cole francaise d'Extrême Orient*, Tome iv, 1904, with an able introduction. An essay by the same scholar on the life and date of Vasubandhu is also published in the Journal of the *Royal Asiatic Society of Great Britain and Ireland*, for 1905 pp. 44-53. In this essay Dr. Takakusu assigns to Paramârtha a period from A. D. 499 to 569 and to Vasubandhu from A. D. 420 to 500. Vincent Smith, in his *Early History of India*, 3rd edition, appendix N, p. 328, carries it back still further by about 200 years. But we are here concerned not so much with the dates of these Buddhistic authors as with the text of the *Sâṅkhya-Kârikâs*; and looking at the question from this point of view we find that the commentary translated into Chinese is not the same as that of Gauḍapâda. The point is noticed by Dr. Takakusu in his introduction, where he has also given a tabular statement of the difference between the Chinese and Gauḍapâda's commentary. The general trend of the two commentaries is of course the same, but the text of each is evidently different. The Chinese commentary is no doubt a translation of an old Sanskṛit commentary on the *Kârikâs*; but what this Sanskṛit commentary was, is still an unsolved question. In the Deccan College Library, there is a Ms. of the *Sâṅkhya-Kârikâs* (No. 197 of 1871-72) wherein the commentary is called Mâṭhara-vṛitti. This is more complete than Gauḍapâda's *Bhâshya* and it agrees more closely with the commentary translated into Chinese. But on comparing it with the Chinese version in some important places, I find that two cannot be taken as identical. There is a third, and I might say a much more recent commentary on the *Kârikâs* viz., the *Sâṅkhya-tattva-Kaumudî* by Vâchaspati-mishra, an edition of which with a gloss. has been published by Paṇdit Jyeshṭhârâm at the Nirṇaya Sâgar Press in Bombay.

Now turning to the text of the *Kârikâs* as represented in these editions we find that Wilson's edition contains 72 *Kârikâs* in the Ârya metre; and in examining the number and the contents of the *Kârikâs*, I have taken this edition as the standard for comparison. Of these 72 verses the last three tell us how Kapila taught the doctrine to Âsuri (verse 70) and the latter to Pañchashikha, and how from him through succession of teacher and pupil it was learnt by Ishvarakṛishṇa (verse 71), who finally condensed it (verse 72) from *Shashṭhi-tantra* into 70 *Âryâs* or verses. This evidently means that the main or the doctrinal part of the book (*i. e.* excluding the concluding three *Âryâs*) consisted of 70 verses. But on the other hand, if we exclude the three concluding *Âryâs* there remian only 69. Again if we look to Gauḍapâda's *Bhâshya* we find that it ends with the 69th verse, and yet Gauḍapâda at the end expressly states that there are 70 *Kârikâs* in the text. Wilson has noticed this discrepancy. In his comment on the 72nd *Ârya* he says " we have here in the text reference to seventy stanzas, as comprising the doctrinal part of the Sâṅkhya. In fact, however, there are but sixty-nine." It might be contended that the number seventy may be made up by adding the first of the three concluding *Âryâs* to the previous sixty-nine. But, observes Wilson, that if the first of the last three stanzas containing the notice of Kapila (verse 70) were to be included in the enumeration, it might fairly be asked "why should not the next stanza at least (verse 71), making mention of the reputed author (Ishvarakṛishṇa), be also comprehended in it, when there would be seventy-one verses ? The Scholiasts offer no explanation of this difficulty." Nor does Wilson give any.

In the Chinese translation by Paramârtha there are 71 *Kârikâs* only. But the sixty-third *Kârikâ* in Wilson's Edition is omitted in the Chinese. This omission is evidently an error; for, as observed by Dr. Takakusu, the verse is found in *Sânkhya-Sûtras* iii. 73, and also in the *Bhâṣhya* of Gauḍapâda. If we supply the omission, the Chinese version will have the same 72 *Kârikâs* as in Wilson's edition. Another remarkable fact is that the *Kârikâs* are called *Âryâ-saptati* in Sanskrit, and are denominated *Suvarṇa-saptati* (the gold-seventy) in the Chinese. This further confirms the statement made above regarding the existence of seventy stanzas in the doctrinal part of the work.

The Deccan College Library manuscript of the Mâṭhara-vṛitti apparently contains 69 verses only. But the verses numbered 57, 58 and 59 in Wilson's edition are wanting in the body of the Ms. They are, however, written on the margin together with the *Vṛitti*, evidently by some one who corrected the manuscript. This Ms. must, therefore, be supposed to contain the same 72 verses as in Wilson's edition.

The *Sânkhya-Tattva-Kaumudî*, a commentary on the Kârikâs by Vâchaspati-mishra, as printed in the Nirṇaya Sâgar Press, Bombay, also contains the same seventy-two verses.

We thus see that there are in all 72 verses in all the copies of the *Sânkhya-Kârikâs* now available. The last three of these give us only the line of succession of teacher and pupil or *gura-paramparâ* as it is called. Thu the doctrinal part of the book at present contains 69 verses only, whereas verse 72 expressly tells us that there are 70 verses in the work. No satisfactory solution of this difficulty has to my knowledge yet been published.

Let us, therefore, see if we can find a clue to this missing verse. The sixty-first *Kârikâ* and Gauḍapâda's *Bhâshya* upon it in Wilson's edition run as follows—

प्रकृतेः सुकुमारतरं न किंचिदस्तीति मे मतिर्भवति ।
या दृष्टास्मीति पुनर्न दर्शनमुपैति पुरुषस्य ॥ ६१ ॥

लोके प्रकृतेः सुकुमारतरं न किंचिदस्तीत्येवं मे मतिर्भवति येन परार्थ एवं मतिरुत्पन्ना कस्मादहमनेन पुरुषेण दृष्टास्मीत्यस्य पुंसो पुनर्दर्शनं नोपैति पुरुषस्यादर्शनमुपयातीत्यर्थः । तत्र सुकुमारतरं वर्णयति । ईश्वरं कारणं ब्रुवते । अजो(?ज्ञो) जन्तुरनीशोऽयमात्मनः सुखदुःखयोरीश्वरप्रेरितो गच्छेत् स्वर्गं नरकमेव वा । अपरे स्वभावकारणिकां ब्रुवते । केन शुक्लीकृता हंसा मयूराः केन चित्रिताः । स्वभावेनैवेति । अत्र सांख्याचार्या आहुः । निर्गुणत्वादीश्वरस्य कथं सगुणतः प्रजाः जायेरन् कथं वा पुरुषान्निर्गुणादेव । तस्मात् प्रकृतेर्युज्यते । तथा शुक्लेभ्यस्तन्तुभ्य शुक्ल एव पटो भवति कृष्णेभ्यः कृष्ण एवेति । एवं त्रिगुणात् प्रधानात् त्रयो लोकास्त्रिगुणाः समुत्पन्ना इति गम्यते । निर्गुण ईश्वरः सगुणानां लोकानां तस्मादुत्पत्तिरयुक्तेति । अनेन पुरुषो व्याख्यातः । तथा केषांचित् कालः कारणमित्युक्तं च । कालः पञ्चास्ति (? पचति) भूतानि कालः संहरते जगत् । कालः सुप्तेषु जागर्ति कालो हि दुरतिक्रमः ॥ व्यक्तम (?क्ता) व्यक्तपुरुषास्त्रयः पदार्थास्तेन कालोऽन्तर्भूतोऽस्ति स व्यक्तः । सर्वकर्तृत्वात् कालस्यापि प्रधानमेव कारणं स्वा (?स्व) भावोऽप्यत्रैव लीनः तस्मात् कालो न कारणं नापि स्वभाव इति । तस्मात् प्रकृतिरेव कारणं न प्रकृतेः कारणान्तरमस्तीति । न पुनर्दर्शनमुपयाति पुरुषस्य । अतः प्रकृतेः सुकुमारतरं सुभोग्यतरं न किंचिदीश्वरादि कारणमस्तीति मे मतिर्भवति ।

Here the 61st verse clearly states that there is nothing more 'delicate' or 'subtle' (**सुकुमारतर**) than

Prakṛiti, evidently meaning that according to the Sâṅkhya philosophy *Prakṛiti* is the final cause of the visible world and that there is no other finer, or subtler or ulterior cause. The comparative degree (सुकुमारतर), here used, shows that the writer of the *Kârikâs* must have in his mind some other ulterior causes of the world mentioned by other philosophers. What could these be ? There is no stanza in the present text which would elucidate this point. But if we read Gauḍapâda's *Bhâṣhya*, the point is cleared up by mentioning four subtler causes advocated by others, namely, *Ishvara*, *Puruṣha*, *Kâla*, (time) and *Svabhâva* (nature). The first two of these being *nirguṇa* cannot be the cause of *saguṇa* world, and the last two being *vyakta* (visible) cannot precede the *avyakta* (*Prakṛiti*). So the *Bhâṣhya-kâra* re-iterates the Sâṅkhya conclusion that there is no finer or subtler cause of the visible world than the *Prakṛiti* ; and then proceeds to the examination of the next, that is, the 62nd verse in Wilson's edition.

Now as I was reading this *Bhâṣhya*, it struck me that all this discussion about several ulterior causes of the world could not have been inserted in the *Bhâṣhya* by Gauḍapâda on his own responsibility. On the contrary he introduces it by the phrase, तत्र सुकुमारतरं वर्णयति, which means, (he) now describes or explains (what is meant by) सुकुमारतर. The third personal verb वर्णयति cannot refer to the Bhâṣhya-kâra or commentator. Again the enunciation of the different doctrines regarding the ulterior cause of the world is introduced by such words as ईश्वरं कारणं ब्रुवते, अपरे स्वभावकारणिकां ब्रुवते, तथा केषांचित् कालः कारणमिति, नापि स्वभाव इति, and the reply is introduced by the phrase, अत्र सांख्याचार्या आहुः। (See the underlined words in the extract from the Bhâṣhya given above.) These phrases, coming as they do after the phrase तत्र सुकुमारतरं वर्णयति, look like excerpts, or *pratîkas* as they are called in Sanskṛit, from the text on which Gauḍapâda is commenting. In other words Gauḍapâda here seems to have before him a verse in the text which explained why *Prakṛiti* was called सुकुमारतर in the 61st stanza. It may be noted that in commenting on the 27th *Kârikâ* Gauḍapâda while explaining how the variety in the world is produced by गुणपरिणाम alone, has discussed the same different causes, viz., *Ishvara*, *Svabhâva*, *Purusha*, etc. But there Gauḍapâda uses no such phrases as सुकुमारतरं वर्णयति, nor is there to be found a systematic statement and refutation of different opinions as here. This leads me to conclude that in the original *Kârikâs* there must have been a verse following the 61st and explaining why *Prakṛiti* was therein called सुकुमारतर. Taking my cue from the underlined words in the *Bhâshya* above given I would, therefore, reconstruct and restore the now missing *Âryâ* as follows:—

कारणमीश्वरमेके पुरुषं कालं परे स्वभावं वा।
प्रजाः कथं निर्गुणतो व्यक्तः कालः स्वभावश्च॥

The first half of this Ârya would thus give the four causes (subtler than *Prakṛiti*) mentioned by others, and the second half would contain their refutation in brief; and then तत्र सुकुमारतरं वर्णयति in Gauḍapâda's Bhâṣhya would be a fitting introduction to it. It may be noticed that पुरुष, here mentioned is निर्गुण पुरुष and that the reply प्रजाः कथं निर्गुणतः holds good equally in the case of ईश्वर and पुरुष; while the reply that व्यक्त cannot be the cause of the अव्यक्त (*Prakṛiti*) applies both to काल and स्वभाव mentioned in the first half of the proposed Ârya.

The peculiarity of the Sâṅkhya as distinguished from other systems of philosophy is that in their search for a final cause of the visible world the Sâṅkhyas never go beyond *Prakṛiti*. They recognise neither *Ishvara* nor anything else as standing behind and controlling the *Prakṛiti* (Cf. Bhagavad-gîtâ ix. 10.) The *Sâṅkhya-pravachana-Sûtras* i, 92 & v, 2ff. clearly state that Ishvara cannot be proved to be the final cause, and the *Sâṅkhya-Kârikâs* would be incomplete without a similar statement. But the doctrine is not contained in the 69 *Kârikâs* forming the doctrinal part of the present text. Unless, therefore, we supply a verse like the above, the doctrinal part of the *Sâṅkyha Kârikâs* would be seriously defective, not to mention that one Ârya would be wanting to complete the number of 70 *Âryâs* said to be originally comprised in the book.

I have tried to restore the lost verse from the *Bhâṣhya* of Gauḍapâda. But it may fairly be asked if the part of the Bhâṣhya relied upon is genuine. For this purpose we may refer to Dr. Takakusu's French translation of the Chinese version of the Kârikâs and the commentary referred to above. The Chinese commentary on the 61st verse is more complete than Gauḍapâda's *Bhâṣhya*. It mentions more fully than Gauḍapâda the four causes, viz. *Ishvara*, *Purusha*, *Kâla* (time) and *Svabhâva* (nature), which some believe to be subtler than *Prakṛiti*, and refutes them one by one by the same arguments as used by Gauḍapâda. It is very difficult to judge what exact words were used in the Sanskrit commentary which was rendered into Chinese. In the French translation the adversary's objection is introduced by the words "On pourra dire" (one might say, corresponding to the एके ब्रुवते and अपरे ब्रुवते in Gauḍapâda's Bhâṣhya) and to the refutation are prefixed the words "Respondent a tous nous disons" (Replying to all this we say, corresponding to अत्र सांख्याचार्या आहु: of Gauḍapâda'). It seems clear, therefore, that the commentary which was rendered into Chinese by Paramârtha in the early part of the sixth century A. D. was in substantial agreement with the *Bhâṣhya* of Gauḍapâda.

I have stated above that the manuscript of Mâṭharavṛitti in the Deccan College Library agrees more with the Chinese commentary than does the *Bhâṣhya* of Gauḍapâda. The Mss. of this vṛitti are very scarce.

The Deccan College Ms. gives the commentary on the 61st verse thus—

प्रकृतेः सुकुमारतरं न किंचिदस्तीति मे मतिर्भवति। एवं बुद्धिरुत्पन्ना। या दृष्टास्मीति पुनर्न दर्शनमुपैति पुरुषस्य। यथेह स्नुषा व्रीडया स्वगृहांतः प्रविशति। अतो ब्रवीतीयं विशिष्टतरेति। एवमियं प्रकृतिः सुकुमारतरा। तस्मात् प्रकृतेः सुकुमारतरं नान्यदस्तीति। एवं पुरुषस्य परमात्मनोऽपि मतिरुत्पन्नास्ति। मे मतिर्भवतीति पुरुष आत्मानं ब्रवीति। तत्र सुकुमारतरत्वं वर्णयति। ईश्वरः कारणमिति केचिदाचार्या ब्रुवते। उक्तं च। अज्ञो जंतुरनीशोऽयमात्मनः सुखदुःखयोः। ईश्वरप्रेरितो गच्छेत्स्वर्गं नरकमेव वा॥ वेदवादिनः पुनरित्थं कारणमाहुः। पुरुष एवेदं सर्वमित्यतः पुरुषं कारणमाहुः। अपरे स्वभावमाहुः। स्वभावं कारणमिति। तथाहि। केन शुक्लीकृता हंसाः शुकाश्च हरितीकृताः। मयुराश्चित्रिता येन स नो वृत्तिं विधास्यति। अत्र सांख्या वदंति। ईश्वरः कारणं न भवति। कस्मात्। निर्गुणत्वात्। इमाश्च सगुणाः प्रजाः। सत्वरजस्तमांसि त्रयो गुणाः। ते च प्रजासु सन्ति। तांश्च गुणान् दृष्ट्वा साधयामः प्रकृतेरिमाः प्रजाः समुत्पन्नाः। यदीश्वरः कारणं स्यात् ततो निर्गुणादीश्वरान्निर्गुणा एव प्रजाः स्युः। न चैवम्। तस्मादीश्वरः कारणं न भवति। एवं पुरुषोऽपि द्रष्टव्यः। स्वभावो नाम न कश्चित्पदार्थोऽस्ति यतः प्रजानामुत्पत्तिसंगतिः स्यात्। तस्माद्यो ब्रूते स्वभावः कारणमिति तन्मिथ्या। केचित्कालं कारणतया वर्णयन्ति। कालः सृजति भूतानि कालः संहरते प्रजाः। कालः सुप्तेषु जागर्ति तस्मात्कालः कारणम्। तदपि सांख्या निराकुरुते। कालो नाम न कश्चित्पदार्थोऽस्ति। व्यक्तमव्यक्तं पुरुष इति त्रय एव पदार्थाः। तत्र कालोऽन्तर्भूतः। एवं प्रधानं हित्वा नास्त्यन्यत्कारणम्। तत्प्रधानावगमं प्रति यदा पुरुषस्य सम्यग्ज्ञानमुत्पद्यते तदा तेन ज्ञानेन दृष्टा प्रकृतिः पुरुषसंगान्निवर्तते। तस्यां च निवृत्तायां मोक्षं गच्छति। एवमीश्वरादीन्यकारणनि। सुकुमारतर इत्येतेन वाक्यशेषः कृतः। यस्मात्सुकुमारतरं प्रधानं तस्मादुच्यते सुकुमारतरं न किंचिदस्तीति। मे मतिर्भवतीति पुरुषस्य।

अत्राह । सा दृष्टा पुरुषेण कथं निवर्तते प्रकृतिः । अत्रोच्यते । या दृष्टास्मीति पुनर्न दर्शनमुपैति पुरुषस्य । यथा काचित् कुलस्त्री साध्वी स्वगृहद्वारि स्थिता पुरुषेण सहसैवागतेन दृष्टा सहसैव व्रीडमाना त्वरितं गृहं प्रविष्टा । सा एवं मत्वा दृष्टाहमनेनेति न पुनर्दर्शनमुपैति पुरुषस्य । एवं प्रकृतिः परमात्मना पुरुषेण ज्ञानचक्षुषा दृष्टा । सा सव्रीडा न पुनर्दर्शनमुपैति पुरुषस्य । तस्यां च निवृत्तायां पुरुषो मोक्षं गच्छति ।

If we compare this *Vṛitti* with Gauḍapâda's *Bhâshya* on the 61st verse we shall find that both are substantially the same. The phrases तत्र सुकुमारतरत्वं वर्णयति and ईश्वरः कारणमिति etc. occur in both, and even the quotations in support of the adverse theories are almost identical. Still more important are the words सुकुमारतर इत्येतेन वाक्यशेषः कृतः showing that, in the opinion of the *Vṛittikâra* the passage which follows तत्र सुकुमारतरत्वं वर्णयति is explanatory of the statement in the 61st verse, viz., प्रकृतेः सुकुमारतरं न किंचिद् etc. Of course this वाक्यशेष may be supposed to have been made either by the text-writer or by the commentator. But the phrase तत्र सुकुमारतरत्वं वर्णयति precludes the supposition that it is made by the commentator. Here we have, however, a new and different explanation of the phrase मे मतिर्भवति inasmuch as मे is interpreted to mean पुरुषस्य *i. e.* परमात्मनः । This explanation is more or less Vedântic. The Sâṅkhyas do not recognise परमात्मा and it seems to me that this explanation is an interpolation made by some one who was anxious to interpret the *Kârikâs* consistently with the Vedântic view of परमात्मा. Ishvarakṛishṇa was in my opinion a thorough निरीश्वरवादी, that is, recognising nothing beyond पुरुष and प्रकृति. But it seems that, later, attempts were made to explain his *Kârikâs* consistently with Vedânta, as Vijñâna Bhikṣhu has done in his *Sâṅkhyasâra*. Ishvara-kṛishṇa in the hands of these commentators thus became as Vedântist, and then it was, I think, that the verse originally following the 61st was dropped and the commentary thereon, that is, the old commentary सुकुमारतरं वर्णयति etc. was tacked on to the 61st verse. When this change took place it is impossible to exactly ascertain; but it must have taken place before the *Kârikâs* were rendered into Chinese. It is noticeable in this connection that the *Sâṅkhya-tattva-kaumudî*, a commentary on the *Kârikâs* by Vâchaspati-mishra, does not at all notice this lengthy discussion about the several ulterior causes of the world. Vâchaspati-mishra is believed to have lived in the 12th century after Christ.

To sum up. (1) The *Sâṅkhya-Kârikâs* are known as *Âryâ-Saptati* (see verse 72) in Sanskṛit and *Suvarṇa-Saptati* in Chinese. (2) But our present text contains 72 verses, and when the last three are excluded as giving only the गुरुपरंपरा there remain only 69 verses for the doctrinal part of the book. (3) The *Bhâshya* of Gauḍapâda, the माठरवृत्ति, and the commentary translated into Chinese all contain lengthy and substantially similar discussions on verse 61, explaining why in that verse प्रकृति is called सुकुमारतर. (4) In these discussions there are words which indicate that the discussion must have been originally based on a verse in the text, and is not an exposition given merely by the commentators. And (5) an essential characteristic of the Sâṅkhya doctrine will be wanting in the *Kârikâs* if the work be supposed to consist in its doctrinal part of only 69 verses. If we put all these facts together we are led to infer that originally there was one verse between the 61st and the 62nd (in Wilson's edition) and that if reconstructed from what look like excerpts from it in the old commentaries it would run thus—

कारणमीश्वरमेके पुरुषं कालं परे स्वभावं वा ।
प्रजाः कथं निर्गुणतो व्यक्तः कालः स्वभावश्च ॥

I have stated above that it is an essential part of the Sâṅkhya doctrine not to recognise any cause of the world subtler than *Prakṛiti*: neither *Ishvara*, nor *Purusha*, nor *Kâla* (time), nor again *Svabhâva* (nature); and that the subject is noticed twice in the old commentaries on the Kârikâs, once in explaining the 27th verse and again in the commentary on the 61st verse. It is interesting in this connection to note that the Arab writer Alberuni, quoting from a Sâṅkhya book in the form of a dialogue, dwells upon the same essential doctrine of the Sâṅkhya philosophy (*Vide Alberuni's India*, English trans. Vol. I. pp. 30 and 31, Trubner and Co.). With this independent evidence regarding the characteristic doctrine of the Sâṅkhyas before us, it would certainly be unreasonable to suppose that the doctrine was not mentioned in

Sâṅkhya-Kârikâs, as we shall have to do if the doctrinal part of the text is believed to have originally contained only 69 verses found in the existing editions of the Kârikâs. *Shvetâshvatara Upanishad* vi. 1, it may be finally mentioned, expressly refers to *Svabhâva*, *Kâla* and *Ishvara* as the three possible ulterior causes of the world and naturally declares the last one, viz., *Ishvara*, as being the real cause. The *Sâṅkhyas* reject all these three, and from the remaining two, *Purusha* and *Prakṛiti*, reject *Purusha* as it is *nirguṇa* and fix on *Prakṛiti* alone as the final cause of the visible world. If the *Shvetâshvatara* refers to this discussion, there is no reason why the *Sâṅkhya-Kârikâs* should not have originally contained a similar discussion of the subject.

CHALDEAN

AND

INDIAN VEDAS.

(1917)

CHALDEAN

AND

INDIAN VEDAS.

One of the most important events of the latter half of the nineteenth century is the discovery of the Chaldean literature as embodied in the cuneiform inscriptions excavated in Mesopotamia and deciphered with great skill, ingenuity, and perseverance by European scholars. These ancient records conclusively show that the country at the mouth of the Euphrates was, so far back as 5000 B. C., colonised by a people of the Turanian race who went there by sea from some distant province, presumably situated in Northern Asia. These people not only developed a civilization of their own in Mesopotamia, but what is to the point, have left there a record of their religious beliefs and culture in the form of brick-inscriptions, which M. Lenormant has aptly described as the Chaldean Veda.

* A lecture on this subject was delivered by the late Lok. B. G. Tilak in the hall of the Bombay Presidency Association Rooms at the Appollo Bunder, Bombay, on 6 December 1904, in connection with the Graduates' Association Lecture Series, under the Presidentship of Mr. K. R. Kama; while this article was contributed by him to the *Bhaṇḍārkar Commemoration Volume* with some additions up to date nearly 13 years later i. e. in July 1917

This ancient civilization at the mouth of the Tigris and Euphrates gradually spread northwards and was the parent of the Assyrian civilization which flourished about 2000 years before Christ. It is believed that the Hindus came in contact with Assyrians after this date, and as a natural result of this intercourse Hindu culture was largely influenced by the Assyrian. Thus Rudolph von Ihering, starting with the theory that the original Aryan home was in an uncultivated mountain district in Central Asia, has, in his work on the *Evolution of the Aryans* (Eng. trans. by Drucker, 1897, pp. 11, 223–4), come to the conclusion that the Aryans were originally a nomadic race unacquainted with agriculture, canals, navigation, stone-houses, working in metals, money transactions, alphabet and such other elements of higher civilization, all of which they subsequently borrowed from the Babylonians. But this conclusion is not accepted by other scholars, who think that von Ihering has gone too far in the matter. It is, however, still believed that in the matter of magical charms and formulæ, cosmography, cosmogony, astronomy and chronology the Hindus were more or less indebted to the Babylonians, and that this borrowing was the result of an intercourse between the two races at a date later than 2000 before Christ.* When it was, therefore pointed out that the word *manâ* in the phrase *sacâ manâ hiraṇyayâ*

* For a summary see the article on *Hinduism* in Hastings' Encyclopædia of Religion and Ethics. Vol VI. p. 688f.

(Ṛv. VIII. 78. 2) corresponded with Latin *mina*, the Greek *μνᾶ* and the Phœnician *manah*, and it must therefore have been borrowed by the Indians from the Babylonians, and that, if so, a later date must be assigned to the Ṛigveda. Professor Max Muller declined * to accept the inference and contended that the word might be of Aryan origin and that it might, as interpreted by Sâyaṇa, mean 'ornaments' or 'beautiful appendices'. For Professor Max Muller believed, and rightly, that the Ṛigveda, the oldest of the Vedas, cannot be assigned to a date later than 2000 years before Christ. The learned Orientalist was aware that the word *manâ* was to be found not only in the Babylonean but also in the Accadian tongue. But he seemed not to have realised the importance of this fact ; for in that case, the Accadian being a still older language, it was not necessary to assign a later date to the Ṛigveda even if the word *manâ* (cf. Kanarese and Marâthî *maṇa*, English corruption 'maund') was found to be of foreign origin.

In my *Orion or the Antiquity of the Vedas*, I have shown that Vedic culture or civilization can be carried back as far as, if not further than, 4500 B. C., when the Vernal equinox was in Orion. This makes the Vedic and the Chaldean civilizations almost contemporaneous, and it is not unnatural to expect some intercourse either by

* *India : What can it teach us* ? Edition 1883, pp. 125—26.

land or by sea between the Chaldean and the Vedic races even in those ancient times. No evidence has, however, yet been adduced to prove the existence of an intercourse between these two races in the fourth or fifth millennium before Christ by tracing Vedic words or ideas in the Chaldean tongue, or vice versa. If this evidence is discovered the existing theories about the inter-relation between these two oldest civilizations will have to be greatly modified or revised. But without going so far into the subject I wish in this essay to confine myself to the words and ideas which I have found common to the Chaldean and the Indian Vedas, stating at the same time what little has been done by the previous scholars in this direction.

Professor A. H. Sayce, in his *Hibbert Lectures*, 1887, pages 137–138, observes that in an ancient list of Babylonian clothing *sindhu* is mentioned as a name for muslin or woven cloth, and that it corresponds to the *shadin* of the Old Testament and the *σινδών* of the Greeks. The learned Professor has further stated that this 'muslin' or woven cloth must have been called *sindhu* by the Accadians (Chaldeans), because it was exported from the banks of the Indus (*Sindhu*) to Chaldea in those days (cf. the word *calico* from *Calicut*). He has further noted that this intercourse between the two countries must have been by sea, for had the word passed by land, i. e. through Persia, the initial *s* of the word would have become *h* in Persian mouths.

Here then we have two words: *manâ* borrowed by the Vedic people from the Chaldeans, and *sindhu* borrowed by the Accadians or Chaldeans from the Indians, proving either that these races were neighbours to each other even in Vedic times or that the Chaldean traders had made their way to the mouth of the Indus or to the Western coast of India, each people borrowing from the other according to necessity.

More recently, the excavations made in Asia Minor during the summer of 1907 have brought to light documents which contain the terms of a treaty between the king of Hittites and the king of Mitani (Northern Mesopotamia), of the time of *circa* 1400 before Christ. In these treaties the deities of both these nations are invoked; and among the Mitani gods Hugo Winkler has found the names of Mitra, Varuṇa, Indra and Nâsatyas or Ashvins, one and all of which are Vedic deities. It is, therefore, quite clear that in the fourteenth century B.C. and earlier the rulers of Northern Mesopotamia worshipped Vedic gods. The names of these rulers, it is true, appear to be Persian and not Vedic. But it does not affect the conclusion that Vedic culture and worship were known to and had influenced the Mesopotamian rulers in the fourteenth century before Christ.*

This takes us back to B. C. 1400 or 1500. But we can go still further back and show, that the intercourse

* H. Jacobi's paper in the *Journal of the Royal Asiatic Society* for July 1909, pp. 721–726.

between Chaldea and India existed from a time far anterior to the reign of the Mitanic kings. M. Lenormant has justly observed that while the Aryans worshipped the good and beneficient deities in nature, the Mongolians (to which race the Chaldeans belonged) always tried to propitiate the malevolent spirits ; and hence while *sacrifice* formed the main feature of the Vedic religion, *magic and sorcery* was the main characteristic of the religion of the ancient Chaldeans. Not that there were no Chaldean hymns to the sun-god, but even these were used for magic purposes. *

This shrewd generalisation of the French savant at once enables us to lay our hand upon the Atharva Veda, if we wish to find any parallels to the Chaldean magic formulæ in the Vedic literature. The Vedic religion is very often called the *trayî-dharma* or the religion based only on the three ancient and older Vedas. The Atharva Veda finds no place amongst these three, and there is an old tradition that in point of importance and authority the Atharva does not stand on a par with the Ṛig, the Yajus and the Sâman. Historically speaking it is now further ascertained that the Atharva Veda is much more recent than the three other Vedas. But though comparatively younger, we must at the same time remember that even this recent Veda must be placed at least some twenty-five centuries before Christ in as

* Lenormant's *Chaldean Magic.*, Engl. Trans., pp. 145f, 179 and 319.

much as it is mentioned by name and cited in the Brâhmaṇas and the Upanishads.*

If we therefore discover any names of Chaldean spirits or demons in the Atharva, it could only mean that the magic of the Chaldeans was borrowed, partially at least, by the Vedic people prior to the second millennium before Christ, and that this could not have been done unless the Chaldean people were either the neighbours of Vedic tribes or traded with them even in those ancient days.

Now let us take for comparison Atharva Veda v. 13. It is a hymn against snake poison; and verses 6,7,8 and 10 therein (omitting the accents) run as follows—

असितस्य तैमातस्य बभ्रोरपोदकस्य च ।
सात्रासाहस्याहं मन्योरव ज्यामिव धन्वनो वि मुञ्चामि रथाँ इव ॥६॥
आलिगी च विलिगी च पिता च माता च ।
विद्म वः सर्वतो बन्ध्वरसाः किं करिष्यथ ॥ ७ ॥
उरुगूलाया दुहिता जाता दास्यसिक्न्या (*v. l.* क्नथाः)।
प्रतङ्कं दद्रुषीणां सर्वासामरसं विषम् ॥ ८ ॥
... ...
ताबुवं न ताबुवं न घेत्त्वमासि ताबुवम् ।
ताबुवेनारसं विषम् ॥ १० ॥

The verses have been translated into English by Bloomfield, † Whitney, Griffith and other scholars; but

---

* Bloomfield's Introduction to Atharva Veda in S. B. E. Vol. XLII.

† In the S. B. E. series vol. XLII, p. 28—

I release thee from the fury of the black serpent, the

none of them has attempted to explain the derivation and meaning of the words printed in black in the original and italics in the translation. Their very sound betrays, to a Sanskṛit reader, their foreign origin. But hitherto not only commentators but even translators have failed to explain their true import or origin. The word *Taimâta* again occurs in Atharva Veda V. 18. 4; while *Âligî*, *Viligî* and *Urugûlâ* do not again occur in the Atharva Veda. According to *Kaushika Sûtras* these hymns are recited while performing certain manœuvres in the process of removing the snake poison. But the *Sûtras* do not give any information regarding the origin of the above mentioned words. Griffith says that *Taimâta* and *Apodaka* (waterless) are some unidentifiable

---

Taimâta the brown serpent, the poison that is not fluid, the all-conquering, as the bow-string (is loosened) from the bow, as chariots (from horses). 6.

Both *Âligî* and *Viligî*, both father and mother, we know your kin everywhere. Deprived of your strength what will ye do? 7.

The daughter of *Urugûlâ*, the evil one born with the black— of all those who have run to their hiding-place, the poison is devoid of force. 8.

....

Tâbuvam (or) not *Tâbuvam*, thou (O serpent) art not *Tâbuvam*. Through *Tâbuvam* thy poison is bereft of force. 10.

varieties of snakes and that *Âligî*, *Viligî*, and *Urugûlâ* similarly indicate some other unknown species of serpents. Whitney considers *Taimâta* as a derivative from *Timâta*, while the word *Urugûlâ* is translated by him as "the broad-knobbed one." *Âligî* and *Viligî* (the father and mother) he does not attempt to explain at all. The word *asiknî*, which means black, suggests that *Urugûlâ* is a word borrowed from black races (cf. *asiknîḥ vishaḥ* in Ṛv. VII, 5. 3). But in the absence of any definite knowledge about the magic and sorcery of the black races, it was impossible to trace the origin of these words. The discovery of the Chaldean literature now supplies us with the means of accurately ascertaining the parentage of some of these words. For instance, the serpent *Taimâta* is, I am sure, no other than the primeval watery dragon *Tiâmat* generally represented as a female but sometimes even as a male monster snake in the Chaldean cosmogonic legends; and the word *Apodaka* in the Vedic text indicates that a land species of the same (as opposed to aquatic) is intended to be coupled with it. Tiâmat is the well-known Chaldean androgynous dragon whose fight with Marduk is the subject of some of the cuneiform tablets* of the creation legends. As regards *Urugûlâ* the word appears as *Urugala* or *Urugula* in the Accadian language. Literally, it means "the great (*gal=gula*)

---

* See Sayce's *Hibbert Lectures* pp. 379–384, and *Chaldea*, Story of Nations Series, Chap. VI, p. 16.

city (*uru*)", but is generally used to denote the great nether world, the abode of the dead — a place visited by Istar in her search for her lover Dumuzi or Tamuz. * Personified, it means the deity of the nether world, and a female snake can be fitly described as *Uru-gula's* **daughter.**

I have not been able to trace *Âligî* and *Viligî*, but they evidently appear to be Accadian words, for there is an Assyrian god called *Bil* and *Bil-gi.* At any rate there is no doubt that *Taimâta* and *Urugûlâ* are, in spite of a little difference in spelling, the same as *Tiâmat* and *Urugala* or *Uru-gula* in the Accadian legends, and that these names must have been borrowed by the Vedic people from the Chaldeans, coming in contact with them either as their neighbours or as tradesmen in those early days. When, the old religion of sacrifice was thus tampered with, and hybrid hymns incorporating foreign magical incantations and formulæ were tried to be introduced in the Vedic literature, it was natural that the Veda which contained these incantations should come to be looked upon with scant respect or even with contempt by the orthodox Vedic community, who must then have regarded the Atharva Veda as a novel departure in their religious observances. There are some other words in the Atharva Veda, especially in the poison and witchcraft

* Jensen's *Kosmologie der Babylonier*, pp. 217–222; *Chaldea*, S. N. Series pp. 157, 326f, and 367f.

hymns, which on their face appear to be foreign importations. For instance we may cite *Tâbuvam* * in the hymn we are considering and *Kanaknakam* and *Taudi* in Av., X, 4. Again in the word *Kimidin* which occurs both in the Ṛig and the Atharva Veda, (Ṛv. VII, 104. 23; Av. I. 7. 1) and which indicates goblins, or evil spirits, is derived by Yâska (VI. 119) from *kim idânîm* (what now?), and explained by observing that these ghosts were believed to wander about in search of "what is now happening." This derivation is obviously fanciful; and as the word has a foreign ring, I believe that it is a Chaldean word. For *Ekimmu* and *Dimme* are Accadian words for spirits and *Kimm-dimm* may well have been a word compounded from them to express ghosts of all kinds.

It may be further noted that the Kirâtas, evidently some non-Aryan tribe, are mentioned as dealing in medicinal herbs in Av. X. 4. 14; and Griffith, in a note to Av. V. 13, 5 interprets *Kairâta* as a variety of snake found among the Kirâtas, the barbarous tribes

---

* I think *Tâbuvam* is derived from the Polynesian word *tabu* and means, pertaining to or resulting from *tabu*, i.e. contact with unclean, unholy, or interdicted thing, in which case the disease or evil requires to be treated with sacred incantations. The exorcist asks whether the poison is or is not of *Tâbuvam* character. For the use of *tabus* in Babylonia see *Thompson's Semitic Magic.*

who inhabited woods and mountains and lived by hunting (the *Kirrhadae* of Arrian). It is therefore not to be doubted that the magic and witchcraft hymns in the Atharva Veda do contain some foreign words. But we in India have not the means to thoroughly investigate all of them. We have no library in India, much less an Assyrialogist we can refer to or consult, for obtaining the requisite information on these matters. The Mleccha-prasiddhârtha-prâmâṇyâdhikaraṇa in Jaimini's sûtras (i. 3. 10) shows that even the orthodox Mîmâṁsakas would not have hesitated to recognise the foreign origin of such words if they had but been able to ascertain it definitely.

The Bible often refers to Chaldea and Babylonia. But no one ever dreamt that the account of creation and deluge in the Old Testament could have been, in the main, borrowed by the Hebrew priests from Chaldean sources. A great sensation was therefore caused in Europe when the Chaldean cuneiform tablets of the creation legend were discovered, their translation published and the Hebrew word *Têhom*, which is translated as 'deep' or 'waters' in the first verses of Genesis, Chap. I, was found to be no other than Assyrian *Tamtu* or the Chaldean *Tiâmat*. Even so late as 1902, Professor Delitzsch's lectures on *Babel and Bible* (Eng. trans. New York 1903) were received and criticised in the same spirit. But it may now be taken as established that the Biblical stories of creation and deluge together with the institution of sabbath and even the story of the fall of man by the serpent are all of Chaldean origin. It was further pointed out by Professor Delitzch, the well-known Assyriologist, that the word *Jehovah*, God's secret name revealed to Moses, was also of Chaldean origin, and that its real pronunciation was *Yahve*, and not *Jehovah*; and this derivation is now accepted even by the compilers of the present Biblical dictionaries. But the matter does not really end at this point. Jehovah is undoubtedly the same word as the Chaldean *Yahve*. But we have still to inquire whether the word can or cannot be traced further back. And here we derive great help from the Vedic literature. The word *yahu* (Zend, *yazu*), *yahva*, *yahvat* and the feminine forms *yahvî* and *yahvatî* occur several times in the Ṛigveda; and Grassmann derives them from the root *yah* = to hasten or to drive quickly. The Nighaṇṭu also tells us that the word *yaha* means water (Nig. I. 12) or strength (Nig. II. 9); while the adjective *yahva* (Nig. III. 3; Nir. VIII, 8); means 'great.' *Yahva* in this sense is applied in the Ṛigveda to Soma (Ṛv. IX. 75. 1), to Agni (Ṛv. III. 1. 12) and to Indra. (Ṛv. VIII. 13. 24). It is needless to give further quotations. I may only mention that *yahva* in one instance (Ṛv. X. 110. 3) is used in the vocative case, and Agni is there addressed as "O *Yahva!* you are the sacrificer of the gods." This clearly shows that the word was not only familiar to the Vedic sages, but that it was applied by them to their gods to signify their might, power or strength;

and Griffith has translated it by the English word 'Lord' in several places. Besides, in the Vedic Sanskṛit we have several other words derived from the root *yah* and so cognate to *yahva*, viz. *yahu*, *yahvat*, *yahvī* and *yahvatī*. It is not, therefore, unreasonable to conclude that *yahva* was originally a Vedic word, and though Moses may have borrowed it from the Chaldeans, yet the Chaldean tongue, in which the various other cognate forms of the word are wanting, cannot claim it to be originally its own. Like the word *sindhu* the Chaldeans appear to have themselves borrowed it from the Indians in their mutual intercourse at some remote period of antiquity.

We might say the same about the Chaldean word *Apsu*, or *Abzu*. It is written as *Zu ab* and read as *Abzu*. It denotes the primeval chaos or watery abyss, and is represented as the husband of *Tiâmat*. Marduk had therefore to fight with them both to rescue the powers of light from their clutches. Dr. Jensen * has critically examined the various meanings of this word in the Chaldean literature. But it is unnecessary to go into these details; for the word and its denotation are well established in usage. It is the primeval abyss from which the gods of light have to be rescued by Marduk for the benefit of mankind. This conquest of Marduk over *Apsu* and *Tiâmat* is celebrated in a Chaldean Epic which is now available in translation.†

* In his *Kosmologie der Babylonier*, pp. 243-253.

† See Sayce's *Hibbert Lectures*, pp. 379-384; Jensen's

I have shown above that the word *Taimat* occurs in the Atharva Veda, and that it must have been borrowed from the Chaldean. Such is not however the case with *Apsu*, the husband of *Tiâmat*. In the Ṛigveda we have not only the word *apsu* in several places but the main features of the Tiâmat-Marduk struggle are also to be found in the Vṛira-Indra fight so fully described in the Vedas. I have shown elsewhere * that Indra's fight with the Vṛitra was for the release of captive waters, and that after the fight these waters, till then enveloped and hemmed in by Vṛitra, the Vedic Tiâmat, were set free, by Indra, to flow (*sartave*).† For this very reason Indra is described in the Ṛigveda as *Apsu-jit*. ‡ It is usual to explain the compound word

---

Kosmologie der Babylonier, pp. 273-288; also *Chaldea*, S. N. Series, Chap. VI.

* See *Arctic Home in the Vedas*, Chap. IX, pp. 233-296.

† Ṛigveda i. 32. 12. Curiously enough the same phrase occurs in the Chaldean Creation Tablet No. 4, line 140, where Marduk after defeating Tiâmat, is said to have ordered her (Tiâmat's) waters, which were not coming out, to come forth. The line is so rendered by Dr. Budge; but Jensen, following the Hebrew tradition, translates it to mean "ordered the waters not to come forth" (*Kosmologie der Babylonier*, p. 288). Vedic tradition and phraseology both support Dr. Budge's rendering and I prefer it to Jensen's. Prof. Sayce (*Hibbert Lectures*, 1887, p. 383) follows Dr. Budge, and Jastrow (*Babylonia and Assyria*, p. 438) follows Jensen.

‡ Ṛv. VIII 13. 2; VIII. 36. 1; IX. 106. 3.

Apsu-jit by treating its first member as a locative of *ap* = water and translate it as meaning "conqueror *in* waters." But it will be easily seen that in spite of the Vârtika on Pâṇini VI. 3. 18, this is rather a forced construction, and that it is better to take *Apsu* as a word by itself and translate *Apsu-jit* as "conqueror of *Apsu*." The same remark applies to the words *Apsu-ja* and *Apsu-kshit* and the like. It may be further noted that the phrase *apsavam arṇavam* * also occurs in the Ṛigveda, and there, *apsavam*, which is an adjective, evidently means "of or relating to *Apsu*." Similarly the word *apsumat* is also found in the Vedic literature (Ait. Brâh. VII. 7), and it is there applied to Agni. In this word we cannot take *Apsu* as a locative of *ap*; and if we have thus a direct authority for treating *Apsu* as a separate word by itself, there is no reason why we should not take *Apsu* as a word by itself, and not as the locative of *ap*, in such words as *Apsu-jit* and *Apsu-kshit*. *Apsu* taken as a separate word, may be derived either from *ap* = water and *su* = to beget, or from *psu*, which, according to Nig. III. 7, means shape or form. In the latter case *Apsu* would mean a shapeless or formless chaos, which is the meaning assigned to it in the Chaldean literature. Anyhow there is little doubt that *Apsu* in *Apsu-jit* is the same word as the Chaldean *Apsu* or *Abzu* which was conquered by Marduk, the Chaldean Indra. The word is evidently Vedic, but owing to the ignorance of its true significance, the Indian etymologists have treated it as the

* Ṛv. X.55.37.

locative of *ap* in compounds like *apsu-jit*. The light thrown by the Chaldean literature on the subject enables us now to rectify the error and understand *Apsu-jit* in its proper and legitimate sense. Tiâmat was the original Chaldean word for the primeval abyss. But when the Vedic word *Apsu* was borrowed it became necessary to differentiate between the two, and this seems to have been done by making one the husband of the other.

Another Vedic word on which new light is thrown by the Chaldean literature is *uru*. In the Vedic literature the word occurs several times by itself and also in compounds like *uru-krama* (Ṛv. I. 90. 9), *uru-kshaya* (Ṛv. X. 118. 9), *uru-gâya* (Ṛv. I. 154. I) and several others. The word *uru* in these compounds is generally taken as an adjective meaning "wide." Thus *uru-gâya* is translated by "wide-going" and so on. But it seems to me that if we take *uru*, as in the Chaldean, to mean the nether world, the above Vedic words can be better interpreted. In the Ṛigveda *uru gâya* is not only applied to Vishṇu but also to Indra and Soma. Now we know from the Ṛigveda that Vishṇu and Soma are the deities who helped Indra in the conquest of the waters of *Apsu*. All these deities can therefore be aptly described as *uru-gâya*, that is, those who traverse the nether world of waters and conquer, along with Indra, the powers of darkness therein. In other words, we can now take *uru-kshaya* as a synonym for *apsu-kshit* and *uru-krama* as synonymous with *apsu-sad* or *apsu-jit*.

The word *uru* appears to have the same meaning in *uru-lokam* in Ṛv. X. 128. 2. But a still more important word is *Uru-ashi* (Urvashî), the name of a well-known nymph. Yâska* would have us believe that the word *uru* in *Uru-ashi* means a thigh, and there is an etymological myth which tells us that *Uru-ashi* was born from the thigh of Nârâyaṇa.† But all these strange derivations become unnecessary if we take *uru* in *Uru-ashi* to mean the nether world or its waters as in the Chaldean. *Uru-ashi* would then mean a watery nymph or a nymph of the nether world and can thus be properly described as *ap-saras*. There are a few other words in the Ṛigveda on which new light may be thrown by the discovery of Chaldean literature. For instance *sinîvâlî* ‡ looks to me like a foreign word, and *tur-pharitû* in that well-known unintelligible verse (Ṛv. X. 106. 6) also wears a suspicious look. I shall not, therefore, be suprised if that verse is found to contain some words of foreign origin. On the other hand *itu* meaning 'a month' ' in the Chaldean language seems to me to be the same word as the Vedic *ṛitu* meaning 'a season' or 'a month.'

---

* Nir. V. 13. † Cf. Kâlidâsa's Vikramorvashîya, i. 3.

‡ *Sin bubbuli* in Chaldean may mean 'disappearance of the moon'; and *tur-parattu* may mean 'son of waters'. *Sindhu-mâtarau*, sons of ocean, is one of the epithets of Ashvins in the Ṛigveda. The word *sina* appears in Ṛigveda ii. 302. and there it is said to be brought to or presented to Vṛitra. Can

VEDIC CONCEPTION OF THE WORLD.

(From The *ARCTIC HOME IN THE VEDAS.*)

VṚITRA TRIUMPHANT. VṚITRA SLAIN.

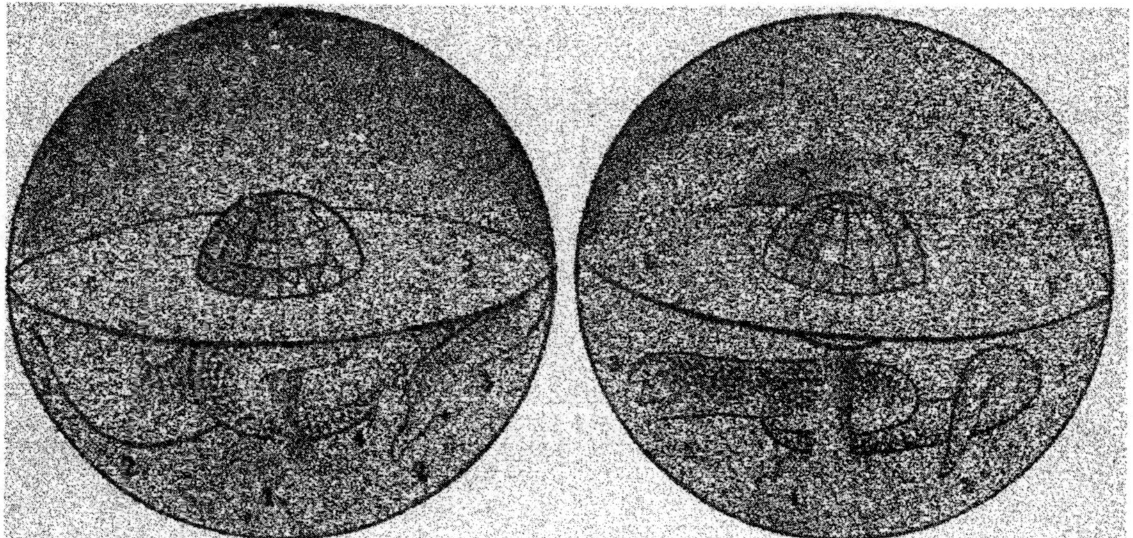

1 Varuna's tree in the Nether World.
2 The cosmic waters (stagnant, Chald. *APSU*) enveloped by Vritra.
3 Vritra as Serpent (Chald. *TIAMAT*).
4 The sun in stagnant waters in the clutches of Vritra.
5 The sunrise mountains.
6 The sunset mountains.
7 The Cosmic waters set free to move in the Heavens.
8 The sun moving with released waters in the Heavens.

BABYLONIAN CONCEPTION OF THE WORLD.

(From JENSEN'S *COSMOLOGIE* Der *BABYLONIER.*)

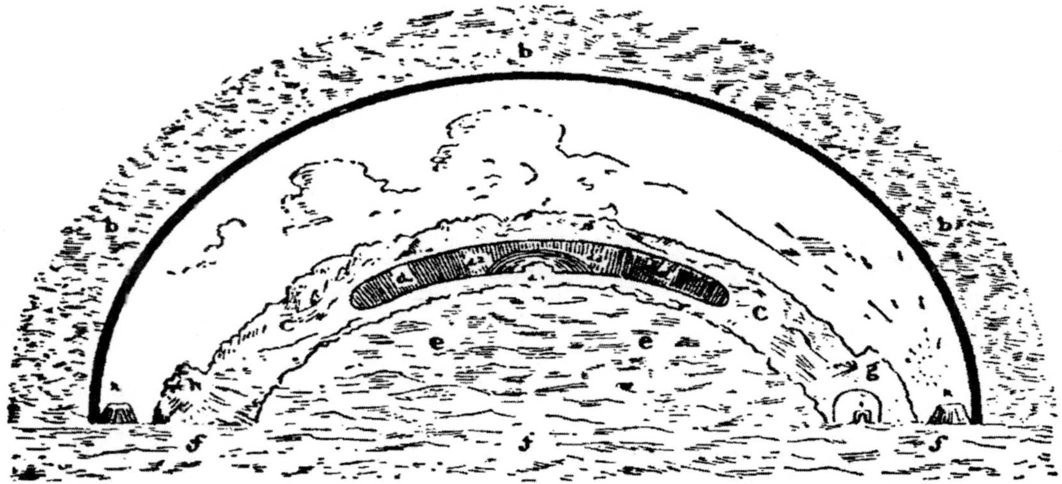

a Inside of the Heavens.
b The Heavenly Ocean.
c The Earth.
d The Kingdom of Death.
  d 1 Palace of Death.
  d 2 The seven walls of Death's Kingdom.
e The prime Hidden place.
f The Ocean (*Apsu*) round the world,
g The sunrise mountain.
h The sunset mountain.
i The meeting-place with the House of Fates.
k The damm of the Heavens.

Lastly I may here mention that we find a very close resemblance between the Chaldean and the Vedic legends regarding the place and movements of cosmic waters, their conquest by the powers of light, viz. by Indra or Marduk, and also between the cosmographic ideas of the two nations, that is those relating to the arrangement of the whole universe, as may be seen by a comparison of the illustrative diagram of "the world according to Babylonish ideas" given by Jensen at the end of his book, and the one given by me in my *Arctic Home in the Vedas* at the end of Chapter IX.* Dr. Jensen has also discussed the sevenfold division of the earth's continents by the Babylonians, and pointed out its resemblance with the Paurâṇic account of the seven continents.† But I think that the parallel can be carried much further; for I have shown elsewhere that this sevenfold division is to be found not only in the Purâṇas but also in the Vedas.‡ It is really interesting to note that there are not only seven Heavens and seven Hells in the Chaldean mythology, but that the serpent Tiâmat killed by Marduk is sometimes represented as having seven heads, while Indra is called

sina here mean the moon? Owing to her waning she may be properly said to be handed over or delivered to Vṛitra, the demon of darkness.

* Compare also Maspero's *Dawn of Civilization*, English translation, Vol. II, pp. 542–543.

† *Kosmologie der Babylonier*, pp. 163–184.

‡ Cf. *Arctic Home*, pp. 340f.

Sapta-han or the "Killer of Seven" in the Vedas,* and the closed watery ocean, the doors of which Indra and Agni opened by their prowess, is described as *sapta-budhna* (seven-bottomed) in Ṛv. VIII. 40. 5. Again there are indications in the ancient Chaldean literature of a dark intercalary winter month and of the sun-hero being affected with a kind of skin disease or lost for a part of the year,† thus corroborating the theory of a common Arctic home for all. But the subject, howsoever interesting it may be, cannot be discussed at the end of this paper. My object was simply to draw the attention of Vedic scholars to the importance of the comparative study of Indian and Chaldean Vedas by pointing out some words which, in my opinion, are common to both, and which fairly establish the case of mutual, and not merely one-sided, indebtedness between the almost contemporaneous Aryan and Turanian people. What effect it may have on the current theories about the inter-relation between the two ancient cultures must be left for the scholars to decide. When two civilizations are contemporaneous it is natural to expect some borrowings from each other; but when both are equally old it is difficult to see why, supposing the borrowing is proved, one of them alone should be considered to have borrowed from the other and that too only in later times.

---

* Ṛv. X. 49. 8.

† In Gilgames and Istar myths. In Ṛv. vii. 100.6 Vishṇu is said to be affected with skin disease (*shipivishṭa*). Compare *Arctic Home*, pp. 330–32. See also Plunket's *Ancient Calendars*, pp. 4 and 14. The intercalary or the thirteenth month was called *se-dir*, the dark month of sowing.

OPINIONS

OF

WELL-KNOWN ASSYRIAN SCHOLARS

ON

"CHALDEAN & INDIAN VEDAS"

(Letter from Dr. A. H. Sayce.)

Queen's College,
Oxford
August 14th, 1919.

Dear Sir,

Very many thanks for your Essay on "Chaldean and Indian Vedas", which has interested me greatly. Your association of *Taimâta* and *Urugûla* with Tiamat and Urugalla is very attractive. With the latter word it would also be possible to compare the Sumerian *usûgal* " the great monster " or "serpent."

The discovery of the existence of an Indo-European language, not of the Iranian but of a purely Indian type, in Asia Minor in the 15th Century B. C. has opened up new vistas in philology. When Asia Minor can be properly excavated it is probable that very important discoveries in this direction are awaiting us.

At any rate it is already clear that the languages of Asia Minor and of the Eastern branch of the Indo-European family influenced one another from a *very early* period, and it is probable that Sumerian also was similarly influenced before the Sumerians descended into Babylonia and founded the civilisation of that country. So we may expect to find much borrowing on both sides.

Believe me to be,

Yours very truly,

A. H. SAYCE.

To

Dr. B. G. TILAK,
Hira Lodge,
60, Talbot Road,
Bayswater,
London. W.

(Letters from Dr. T. G. Pinches.)

(Letter No. 1)

Sippara,
10, Oxford Road,
Kilburn, N. W.
August 13th, 1919.

Dear Prof. Tilak,

Many thanks for your paper "Chaldean and Indian Vedas". As far as I have read it, I find it excellent; and I will write you further on the subject later on.

In accordance with your request, I send you the names of Scholars to whom copies might be sent :—

Prof. Stephen Langdon, Reader in Assyrian, Oxford.

Prof. A. H. Sayce, Queen's College, Oxford.

The Reverend Dr. C. J. Ball, Oxford.

Dr. C. H. W. Johns, St. Catherine's College, Cambridge.

Prof. Farbridge, Manchester.

In all probability I shall think of other names by the time I next write to you.

Yours very truly,
THEOPHILUS G. PINCHES.

To

Prof. B. G. TILAK, &c. &c.
60, Talbot Road,
Bayswater, W.

(Letter No. 2.)

Sippara
10, Oxford Road,
Kilburn N. W.
August 16th, 1919.

Dear Prof. Tilak,

Herewith I enclose a few notes which I have written upon your paper "Chaldean and Indian Vedas." Some of the points upon which I have touched, you may have already considered, but as I may have put them in a different light, you will probably not be sorry to see them stated again.

It seemed to me to be especially important to distinguish between the non-Semitic Sumerian, the Semitic Akkadian, and the Semitic Chaldean, the two former early, the last-named of late date. It is also of importance to quote the correct form of every word treated.

I am sorry to say that I cannot think of the names of any further British Assyriologists. As to the German Assyriologists, I no longer know who among them are my friends, nor do I know which among them may have survived the hardships of the last five years. Among my French colleagues, however, I should suggest the following:—

Monsieur Fr. Thureau-Dangin, Musée du Louvre, Paris.

Monsieur le Prof. V. Scheil, Membre de l'Institut, Paris.

Le Conservateur en Chef, Musêe Guimet, Paris.

Monsieur L. Delaporte, 211, Rue de Paris, Clamart (Seine), France. (A real Scholar, and knows it. Full of scepticism with regard to the work of others, but a good fellow). Notwithstanding that he is old, and much occupied, you might send a copy to the following :—

Sir Henry Howorth, K. C. I. E., 45 Lexham Gardens,
Kensington, W.

I used to know a Barrister, Mr. Parmeshwar Lall who once studied Assyrian. He returned to India, however, some years ago, and I do not know his present address. Notwithstanding that yours is a country of exceedingly great extent, you may have some means of finding out something about him. I think the subject might interest him (if, as I suppose, he be still alive), and it would not be a bad thing if he found himself able to interest himself in Assyriology again.

Trusting that the enclosed notes will be of use to you,

I am,

Yours faithfully,

THEOPHILUS G. PINCHES.

ENCLOSURE.

For the due realization of your thesis, and the disarming of criticism, it is needful to keep the gentilic nomenclature more distinct.

I notice that you use the word *Chaldean* as a synonym of the generally-received *Babylonian.* This I should feel inclined to avoid, owing to the fact, that the Chaldeans do not seem to have entered Babylonia, or at least to have attained prominence until after the time of Moses. This naturally has a bearing upon the name Yahwah (an older form than Yahweh). It is gratifying to me that my theory, that the name of the great God of the Hebrews is now regarded as having penetrated into Babylonian under the form of Yâwa or Yaawa (=Yahawah). My papers dealing with this subject appeared in the *Proceedings* of the Soc. of Bibl. Archæology in 1885 and 1892. Fried Delitzsch can only be right in his contention that Yahwah appears in early Babylonian names if his reading be modified from Yahwe to Yahwa, for a later form can hardly have preceded an early one. From what you say, I should feel inclined to advance the theory, that the Aryan Yahve was adopted by the Babylonians and the Hebrews owing to its likeness to their own (perhaps borrowed) *Yahu* (Bab. *Yau*), 'god', which appears in the bilingual syllabaries as a synonym of the common word *îlu*, with the same meaning.

The date of the use of *Yau* by the Akkadians (Semitic Babylonians) is a little before 2000 B. C., and *Yaawa* (which may also be read *ya-a-pi*), occurs about the same date. *Ya*(*a*)*wa* occurs in certain names of Jews during the period of the later Babylonian kings (6th-5th cent. B. C.). All the names in question are those of Jews which, in the O. T. end in *Jah* or *jahu* (*yah* or *yahu*).

In my opinion, there is little or no connection between the story of the Creation in Genesis and that of the Babylonians. The latter contains no direct statement of the creation of the heavens and the earth; it has no systematic division of the things created into groups and classes, such as is found in Genesis; it has no references to the days of Creation; and no appearance of the Deity as the first and only cause of the existence of things. Other differences are, the polytheism of the Babylonian account, and the fact that it appears to be merely the setting of the legend of Bel and the Dragon, which was composed for the glorification of Merodach, the patron deity of Babylon. As the Babylonian account has no reference to the days of Creation, there is, in that version, no mention of the 7-day week and the sabbath. That the sabbath appears therein I freely admit, notwithstanding that the word is mutilated and incomplete. The Hebrew sabbath, however, was fundamentally different from that of the Babylonians, which designated the lunar festival when our satellite 'rested' at the full on the 15th of the month. It was therefore a monthly sabbath, not a weekly one.

With regard to the Hebrew Creation-story, I will even go so far as to say that, instead of being founded upon that of the Babylonians, it was written to refute it — as a more reasonable statement of the first beginning of things. That the writer of the Hebrew account may have been influenced by the Babylonian idea of the beginning of things is not only possible — it is also probable, but he really sent forth his version to combat what he regarded as the errors and the superstitions of polytheism. Probably, too, he rejected the Babylonian evolution-theory, which, as a strict monotheist, would be distasteful to him.

In connection with this negative theory of mine, it is worthy of note, that there is no mention of Tiawath (Tiamat) in the Hebrew account. Tehom is the Deep (unpersonified). Tiawath (otherwise Tâmthu or Tâwthu) is the ocean both personified and unpersonified. *Apsu* is the Akkadian (Semitic Babylonian) form of the Sumerian *Abzu.* The Akkadians often lengthened the final vowels when they borrowed words, and sharpened *b* into *p.* Another example is *sa-bat*, 'mid-rest', "sabbath", Akkadian *sapattu.* The Heb. s*abbath* is a better reproduction.

Tur purattu would not mean 'son of water', but would be a hybrid, meaning 'the young one of the waterway.' The Sumerian form would be *tur puranunu*, 'the young one of the great waterway' (the Euphrates), in Semitic Babylonian *ṣiḫir Puratti.* As far as I know, however, neither of these equivalent expressions occur.

It ought to be noted that *uru* is a common (or the common) Sumerian word for 'city', and *uru-gal* means 'the great city' as the abode of the many spirits of men who have departed, for 'the dead greatly exceed the living.'

I am afraid that 'the dark intercalary month' does not exist. *Se-dir* is for *Se-gur-tar-dir*, 'the additional (month) of grain-cutting'. It was the sun-god Tammuz who passed the winter-months in the underworld. Merodach descended thither to comfort and bring forth the rebellious gods who had received the grace of his pardon.

EXTRACTS

FROM

A ROUGH NOTE BOOK.

MANDALAY JAIL. **B. G. TILAK**

अंभोजिनी वननिवास विलासमेव ।
हंसस्य हंति नितरां कुपितो विधाता ॥
नत्वस्य दुग्धजलभेद विधौ प्रसिद्धाम् ।
वैदग्ध्य कीर्तिमपहर्तुमसौ समर्थः ॥

Bhartṛihari

ज्ञात्वाधर्ममधर्मं च लोभं चोद्यममाश्रिताः ।
युध्यध्वमनहंकाराः यतो धर्मस्ततो जयः ॥

Mahābhārata.

NOTES FROM

HELLEBRANDT'S "VEDIC MYTHOLOGY"

Vol. I :— Page 57. Pages 468–470. *Sûrya's* daughter and Soma. Ṛig-Veda IX. 113. 3.

cf: Vol. II, Pages 40–42 ; and Vol. III, 125 and 367.

Vol. II :— p. 26 — '*Ushas*' is originally the first dawn of the new Year. (So Ludwig relying on Krichenbauer.) p. 28—*Ekâshṭaka* takes the place of *Ushas* in the Gṛihya ritual as new year's night. T. S. VII 4. 8. 1; अन्हां विधात्री T. S. III. 3. 8. 4. cf : A. V. III. 10. 1–8. M. S. II. 13. 10. Sâyaṇa on Ṛig. V. 62. 2 परिवत्सरे = सत्रांते. The words ऋतूनां विधात्री or अन्हां show that it is the first dawn.

p. 29 — *Ushas* is the first important day of the new year. Shatpath B. VI. 5. 1. 8. And Vashishṭha's Hymn Ṛig. VII. 80 — shows the commencement of Uttarâyaṇa (foot note).

p. 30 — In ancient Rome the ten months' year was followed by an unholy period of time dedicated to the manes (पितरः).

p. 31 — Out of the dark time of the year comes up the new light, return Sûrya, Agni and the disappeared Yajna. This is done by Pitars and नवग्वाः (p. 33).

p. 32.—Chand. III. 19. 3 quoted to show that in the morning, songs were usually recited and hence it is छंदस्वति.

p. 33. —The three dawns refer to colours—white, red and yellow according to Hopkins and to ऋतुs according to Hellebrandt. (Both wrong).

p. 33—It seems, therefore, the singing of hymns by the old Indian Kavis, Aṅgiras and Navagvâs, who the new night found; and broke the cow-stall, is also the custom of other people—to sing the new year's night.

p. 38. After quoting R. V. 1. 92. 4 which says, "that *Ushas* broke out of darkness as cows out of their stall" (cf. Ṛig. IV. 51. 2; V. 45. 1) Hellebrandt observes:—

"Here is a *reminiscence of a foregone time* and of *an earlier Home*, interwoven in the Vedic mythological garb. In India, the light of the dawn lasts only for a short time. Already the Atharva Veda speaks, in contrast with R. V., proportionally only seldom of *Ushas*; (and) the Ritual knows of no special offerings to the (*Ushas*). The picture of the Rocks or stalls, out of which the dawn-light is freed, can (will) have its origin not in the Indian climate but in a land where the cattle is really shut up in strong stalls during winter, as the Vendidad 2. 23 (pakhru-maeṣhn umâ naeshn). Rock cave, Rock stalls must have been far more secure shelter than (artificially) built stall.

In a place where the shortest and the longest day only differ by a *few* hours, there is little reason to celebrate the return of light: in the classical (Sanskṛit) poetry the subject, so far as I know, has attracted little attention. The utterances of the Ṛig-Vedic poets regarding the uprising of the year out of winter night, based on ancient times, which could not be well grounded proportionally in the Vedic times, compared well with the driving out of the cattle from its winter stall for the Spring time".

Agni. (*Volume II. pp. 56–178*).

Page 81. — "If we now turn to Ṛig-Veda: If we have succeeded in proving that the *Uṣhas* denotes not only the dawn-light of any one day, but also of the first day of the new year, whether this begins at the winter solstice or the vernal equinox, then the Vedic poets and families must at least be supposed to be acquainted with renewal of fire at the beginning of the new year. But there are more positive grounds in favour of this assumption. I count amongst them the texts which speak of Agni's living in darkness. It is true that the sun acquires new life every morning and that the faintly burning fire is again revived by the प्रादुष्करण of the Agnihotri ; but this is not sufficient to explain the many turns of Agni's flight, his residence in darkness, his long dwelling in darkness and his release therefrom. Above I have already drawn attention to the contra-distinction between *Devayâna* and *Pitṛiyâna*, which influenced even the Ṛig-Vedic world; and have attempted to show the distinction must be taken to hold not in the local meaning of later times, but in the sense of *Dakshiṇâyana* and *Uttarâyaṇa*. With the end of Devayâna expires the service of the Gods and the periods of manes, the देवशत्रव: (Ṛig. VI. 59. 1.) begins. It may remain undecided whether the whole Pitṛiyâna or only a part of the time, perhaps the last months of the year were assigned to the ritual of the manes; or whether different kinds of measurement (of time) prevailed, one in the Ṛig Veda and another in the classical ritual, which performed the *Vâjpeya* etc. in the autumn. The "darkness" has in southern places less significance than in northern; and the representation (of Agni etc.) as a whole might be an inheritance from previous Home *under other latitudes*; but of the significance of the winter solstice for the Vedic times, there can be no doubt. To this time of Pitṛiyâna, the period of the year belonging to the manes I assign the words, which speak of Agni's residing in darkness or the likes. If one takes, for instance, Ṛig. III. 31,3 ff; IV. 1. 11 ff; 2. 17 ff, and asks whether these passages with their implications and hints about the regaining of the Bull, opening of the Rock Stalls, light and darkness, have not for their basis the production of the new fire rather than the Agnihotra of every day; Ṛig. X. 35, testify—without comparing the readings.............................

Probably the daily Agnihotra was known to the Ṛig-Veda. Therefore it is often difficult to distinguish whether the performance of the daily Agnihotra or of the new fire is indicated; for the course of the day is like that of the year in many things. The Brâhmaṇs (books) see even in the daily Agnihotra a Sûrya-magic, when they state that without the dawn offerings the Sun would not rise; but I would rather think that great divisions (of time – e.g. Ayanas) and turning points have been of influence on the first origin of such views than daily appearance (phenomena).

(pages 136-137) — Oftener we hear of the anxiety of Gods regarding the falling out of the Sun from heavens ; it is the metres and स्वरs with which they support the wavering at the turning point. It cannot be anything else than a reminiscence of an old sun-turning festival which was accompanied with songs.

Agni's Flight.

Page 138 ff — The disappearance of the three brothers of Agni is narrated in T. S. VI 2. 8. 4. "It treats, I think, not of the preparation for a single *Yajna* but of the beginning of the ritual year at the beginning of the *Devayâna* following the *Pitṛiyâna*. The whole narrative has too fixed a character to refer to a single sacrifice." The Indian Mythology has chosen that image of lost or concealed fire, after whom the Gods and the Sages seek, to characterise the end of Devayâna. We can see that from the Ṛig-Veda itself. It speaks of angry Agni (V.2.8). Further the three hymns X.51-53 give the same story. In X. 51. 5 Agni is asked to *make the Paths Deva-*

yâna passable and in X. 124. 1, Agni is said to have lived long in deep darkness (Jyog dirghan). This can be intended only for the new commencement of *Yajna* and by *tamas* refers to sacrificeless and dark time.

Pitṛis called देवशत्रव: in VI 59. 1 and *Indrâgni* are invoked to destroy them, that is, the *Pitṛiyâna* is ended. *Indrâgni* appears here to mean the first new moon in the new year.

Page 145—In VI. 8. 4:—

अपामुपस्थे महिषाऽगृभ्णंत ।

विशोराजानं &c.

Apâm Upasthe– "un Schoose der Wasser" is the phrase used. But in VI 9. 7:— त्वमग्ने तमसि तस्थिवासं । the word '*tamasi*' is used instead, showing that '*Apâm Upasthe*' is synonymous with '*tamasi*'. ('पथो देवयानान्' in X. 51. 5; 'अपामुपस्थे' in VI. 8. 4; and 'तमसि' in VI. 9. 7.) So 'Pitṛiyâna' —contrary of Devayâna— 'Apâm Upastha' and 'Tamas' are synonymous!

Upon this Hellebrandt observes as follows:—

(*Page* 146). "The thought that the sun disappears in waters is, in tropical land, only explainable as referring to rain time (cf: rising and setting of the Sun in ocean). If *tamasi* stands near it, it certainly denotes the darkness of the rainy season; but as I have above distinguished it seems probable that in *tamas* we have inherited a reminiscence of an older time *under different skies* and that it denoted originally the winter solstice. The two lines of thought, run into each other so close, that a clearer separation of the two is not possible; *and the mixing up of the two was made still more possible as the rainy season stands at the beginning of the Dakṣhiṇâyana.*"

(INCOMPLETE)

(Some Important Pencil Jottings.)

(B) Arctic Home in the Vedas.

(*Revision.*)

For revising the book the following new works ought to be consulted :—

1. Dr. Geikie—Ice Age (new Edition).
2. Tarr's Geology (American Work).
3. Hellebrandt Vedic Mythology (Sane's Translation).
4. Plunket—Ancient Calendar (Dark Winter Month).
5. Strickland's Slavonic Folk Lore 8 Vols.
6. Dr. Wallace—Island Life and Wonderful Century.
7. Man and Glacial Period by G. F. Wright, (In Scient. Series—No. LXXII).
8. The Polar Aurora by A. Anjot, (In Scient. Series—No. LXXXI).
9. Dr. Wollarton — Prehistoric Antiquities of N. Europe.
10. Logan's Ancient Indian Stones.
11. Dawn of Civilization (translated from the French). (Plate Picture of World according to Chaldean Mythology).
12. Warren's Homeric Age.
13. Leigh Hunt's Mosaic Exodus.

The following Pamphlets and Articles.:—

1. Encyclopædia Britannica (Egyptology, Geology).
2. Dharma—a review.

3. Indian Review (Madras).

4. Raṅgâchârya's letter.

5. Chaldean Mythology (on release of Waters and on waters being released by Moon).

6. Reference from Sat. Brâhmaṇa (to be inserted).

7. Gopath Brâhmaṇa on गवामयनम्.

8. Pâvgi's Book and Essay.

9. Chaldean Myth of Tamuz (Prof. Sayce, Hibbert Lecture).

10. वारुणो वृक्षः Afra. Chittra दानुचित्राः

11, Bloomfield's Articles on नवग्वा in Am. O. S.

12. Mittra Worship.

13. Thebaut's " Indian Thinker ".

14. Satya Vrata Sama. त्रयी परिचय.

15. The Seven (open Court); also Warren's Review.

16. Weber's Omina et Portenta.

17. Shrauta Bhûmi by M. R. Athavale.

18. Mâhâbhârata reference of 7 Kṛittikâs' (कृत्तिका)'s and story of deluge.

19. Gavâmayana by R. Sham Shâstri, B.A., (Mysore Library).

(C) Orion or Researches into the Antiquity of the Vedas.

—o—

(*Recast with additions.*)

This book requires almost to be re-written in order to incorporate new materials made available since 1893 and new results :—

1. Review of Bulher in Indian Antiquary.
2. Review by Whitney in Am. O. S.
3. Dr. Jacobi's Essay in Indian Antiquary.
4. Barth's Review.
5. Bloomfield's Review (Mahratta).
6. Weber's Naksatras (to be translated).
7. Bârhaspatya—Vedânga Jyotisha.
8. Sudhâkar's Vedânga Jyotisha.
9. Sudhâkar's Dingmimânsâ.
10. S. B. Diksit's कृत्तिका: in Ind. Antiquary.
11. Ketkar's Essay on Jupiter (Royal A. Society's Journal,) with comparison from Chaldean Calendar.
12. A letter in "Maratha" on कृत्तिका with quotations from Atharva Veda, also the name for विशाखा (विचृतौ).

13. Reference from Mahâbhârata on मघा being the asterism of पितृ's; and also अभिजित्तु दिवंगता etc.,

14. Plunket's Ancient Calendars.

15. Dionysus by Robert Brown.

16. Chaldean Astronomy by Robert Brown.

17. The Kali Yuga (Madras In. Review.)

18. Thebaut's Criticism (and also Oldenburg's).

19. Nârâyan Iyengar's Aryan Mythology.

20. Whitney's Nakṣhatras.

21. Max Muller's Essay (Preface to Ṛig-Veda).

22. Prehistoric Antiquities of the Aryan Peoples.

23. मघासु गावो हन्यंते (Note to be written).

24. Shâlva Sûtras (Pandit Series).

25. Jacobi's Antiquity of Vedic Culture.

26. S. B. Diksit's "भारतीय ज्योतिष".

27. Map and Chatterji's Book.

28. Fadke's Essay (Kalyan).

New Chapters.

I. *On due east.*

After "the Kṛittikâs". About Kṛittikâs being due east. A refutation of Sudhâkar's "दिङ्मीमांसा." S. B. Diksit's view.

II. *On the vernal equinox.*

or

The beginning of the Zodiac.

After "Orion and his belt". About अश्विनी or Aries being the first asterism or sign. The reason why, as given by Plunket from Chaldean Astronomy.

Additions.

The story of Orion from Brown's Dionysus. Jacobi's arguments re "Kṛittikâs". Vedic Texts quoted by Jacobi—Last Chapter.

(D) Chaldean and Indian Vedas.

1. Lenormant Chaldean Magic.
2. Smith's Assyrian Texts.
3. King's Chaldea.
4. Prof. Sayce's Works (Hibbert Lectures and other Babylonian literature).
5. Prof. Rawlinson's Ancient Monarchies.
6. Bable and Bible.
7. By paths of Bible knowledge Series.
8. Dr. Pinches' Works.
9. Dawn of Civilisation (Maspero's).
10. Story of Nations Series (Chaldea, Assyria and Egypt).
11. Early Babylonian History (New York, 1900).

(E) Books Projected or Suggested.

(*Syllabus for future work*)

---

1. History of Hindu Religion — Vedic, Shrauta, Upanishads, Epic, Pâuranic, Darshanas, Bhakti, Prehistoric—Other religions, Conclusion.
2. Indian Nationalism, (the story of or the aspects or phases of).
3. Pre-Epic History of India.
4. The Shâṅkara Darshana (Indian Monism).
5. Provincial Administration.
6. Hindu Law.
7. Principles of Infinitesimal Calculus.
8. Bhagvat-Gîtâ-Rahasya—Ethics.
9. Life of Shivâji.
10. Chaldea and India.

(POLITICAL.)

Chapters :—

(1) Introductory.

(2) Vernacracy—Châturvarṇya.

(3) Hindu State and Empire.

(4) Buddhism, Shakas and Renovessance.

(5) Mahomedan Conquest and Empire.

(6) Break up—Marâtthâs, Sikhs etc.

(7) British Conquest,

(8) Government by the Crown (Constitution).

(9) Consolidation.

(10) Bureaucracy—its ideals.
(Comparison of Spanish, Austrian and Russian Bureaucracies).

(11) Progress (two opposite views).

(12) Reconciliation.

(F) The Golden Rule as found IN The Maha Bharata.

" यो यथा वर्तते यस्मिन् तस्मिन्नेव प्रवर्तयन् ।
नाऽधर्मं समवाप्नोति नचाश्रेयश्च विंदति ॥ (अंबोपाख्यान)
उद्योगपर्व; १७९; ३०. p. 413.

Positive aspect of the Rule

" तस्माद्धर्म प्रधानेन भवितव्यं यतात्मना ।
तथा च सर्वभूतेषु वर्तितव्यं यथाऽत्मनि ॥ said by विदुर;
शां. १६७; ९. p. 397.

Negative aspect of the Rule

" यदन्यैर्विहितं नेच्छेदात्मनः कर्मपूरुषः !
न तत्परेषु कुर्वीत जानन्नप्रियमात्मनः ॥ १९ ॥

" जीवितं यः स्वयं चेच्छेत्कथं सोन्यं प्रघातयेत् ।
यद्यदात्मनि चेच्छेत तत्परस्यापि चिंतयेत् ॥ २१ ॥
शां. २५८; १९–२१. p. 595.

" यावान्नात्मनि वेदात्मा तावान्नात्मा परात्मनि ।
य एवं सततं वेद सोमृतत्वाय कल्पते ॥
शां. २३८; २२ p. 563.

Compare with—

सर्वभूतस्थमात्मानं सर्वभूतानि चात्मनि ।
ईक्षते योगयुक्तात्मा सर्वत्र समदर्शनः ॥ २९ ॥ गीता अ. ६.

NOTE THIS PASSAGE:—

(said by बृहस्पति)

" आत्मौपम्येन पुरुषः प्रमाणमधिगच्छति ।
न तत् परस्य संदध्यात् प्रतिकूलं यदात्मनः ।
एष संक्षेपतो धर्मः कामादन्यः प्रवर्तते ॥ अनु. p. 361.

"अथापरः प्रक्रमते परेषु तथापरे प्रक्रमते परस्मिन्।
तथैव तेषूपमा जीवलोके यथा धर्मो निपुणेनोपदिष्टः ॥
अनु. ११३; १०. p. 361.

"यस्मिन् यथा वर्तते यो मनुष्यस्तस्मिन्तथा वर्तितव्यं स धर्मः।
मायाचारो मायया बाधितव्यः साध्वाचारः साधुना प्रत्युपेयः ॥
सत्यानृताध्याय; शां. १०९, २९. p. 259.

"सचेन्निकृत्या युध्येत निकृत्या प्रतियोधयेत्।
अथचेत्धर्मतो युध्येत् धर्मेणैव निवारयेत् ॥
शां. अ. ९५–९. p. 227.

"धर्मेण निधनं श्रेयो न जयः पापकर्मणा।
शां. अ. ९५–१६. p. 227.

"न पापे प्रतिपापः स्यात् साधुरेव सदा भवेत्।
वनपर्व (ब्राह्मणव्याध संवाद). p. 547

"यदन्येषां हितं न स्यादात्मनः कर्म पौरुषम्।
अपत्रपेत वा येन न तत् कुर्यात् कथंचन ॥
शां. १२४. ६६. p. 295

This is said in the definition of शील,

वितृत्योपचरन् वध्यः एष धर्मः सनातनः।
वन. p. 33

निकृत्या निकृति प्रज्ञा हंतव्या इति निश्चयः।
न हि निकृतिकं हत्वा निकृत्या पापमुच्यते ॥
वन. p. 136.

सर्वेषां यः सुहृन्नित्यं सर्वेषां च हिते रतः।
कर्मणा मनसा वाचा स धर्मं वेद जाजले ॥
२६१. ९. p. 601.

ERRATA

Page	Line	Incorrect	Correct.
2	23	Devies	Davis
4	29	Burgen,	Burgess,
6	26	Sicon	Sieu
7	5	"	"
13	1	seasons	solstices
13	24	astronimical	astronomical
14	16	or	nor
29	10	Deivis	Davis
37	12	भीपमर्ष्व	भीष्मपर्व
47	12	amendment	emendations
50	12	pakshas	pakshas by
50	23	चेंदवा	चैंदवा
52	19	एकात्र	एकाग्र
55	16	द्विशस.ति's	द्विसप्तति's
60	17	83	63
67	16	कादश	एकादश
74	20	(b)	B
77	17	error	the error
82	6	शेष	शेषं
86	27	पाद्स्तिशत्तु	पादस्त्रिंशत्तु
87	4	(तर्दितं)	(तर्हितं)
89	25	proposed	propose
99	18	त्रिपद्यायास्ति	त्रिपथायास्त्रि
102	28	to change गण	to change गुण
173	13	कर्मपूरुष: I	कर्मपूरुष: ।